ABHANDLUNGEN FÜR DIE KUNDE
DES MORGENLANDES

Im Auftrag der Deutschen Morgenländischen Gesellschaft
herausgegeben von Florian C. Reiter

Band LXI

2007
Harrassowitz Verlag · Wiesbaden

Florian C. Reiter

Basic Conditions of Taoist Thunder Magic
道教雷法

2007

Harrassowitz Verlag · Wiesbaden

Bibliografische Information der Deutschen Nationalbibliothek
Die Deutsche Nationalbibliothek verzeichnet diese Publikation in der Deutschen
Nationalbibliografie; detaillierte bibliografische Daten sind im Internet
über https://dnb.de abrufbar.

Bibliographic information published by the Deutsche Nationalbibliothek
The Deutsche Nationalbibliothek lists this publication in the Deutsche
Nationalbibliografie; detailed bibliographic data are available on the internet
at https://dnb.de.

For further information about our publishing program consult our
website https://www.harrassowitz-verlag.de

© Deutsche Morgenländische Gesellschaft 2007, 2025
Otto Harrassowitz GmbH Co. KG, Kreuzberger Ring 7c-d,
D-65205 Wiesbaden, produktsicherheit.verlag@harrassowitz.de

This work, including all of its parts, is protected by copyright. Any
use beyond the limits of copyright law without the permission of
the publisher is forbidden and subject to penalty. This applies
particularly to reproductions, translations, microfilms and storage
and processing in electronic systems.
Printed on permanent/durable paper.

Printed in Germany
ISSN 0567-4980
ISBN 978-3-447-05593-2

Contents

Foreword	VII
Introduction	1
The Biography of Yeh Ch'ien-shao	5

Chapter I: Man and Thunder Magic

A First Approach to Thunder Magic	13
Human Nature and Thunder Magic	15
Summoning and Uniting [Thunder Divinities]	19
Assembling the Divine Force	32
The Basis for Writing Amulets	41
Praying for Rain	47
Killing and Crusading, the Martial Side of Thunder Magic	54
The Creative Impetus	62

Chapter II: The Scope of Thunder Magic

The Divine Texts of the Great Methods of the Five Thunders at the Jade Department in the Heaven of Highest Purity: Preface by Wang Wen-ch'ing	69
The Arrangement of the Offices of Thunder and Thunderclaps	74
Method of Consecrating Seals	79
Locations and Departments of the Five Thunders	81
Sacrifice to the Great Divinity of Law and Order	85
The Application of Thunder Breaths	87
Writing out Amulets to Cure Illnesses, Raise Dragons and Produce Rainfall	88
The Establishment of Hells for Evil Demons	96
Altars and Prayers for Rain or Clear Skies	100
The Crusade against Temples and the Eviction of Wicked Spirits	109
Ritual Steps and *Mudrâs*	112
Conditions for the Ritual Transmission and [Spirit] Promotion	117

Abbreviations 121
Bibliography 123
The Sources in the Taoist Canon... 126
Glossary 129

Foreword

Today Taoist Thunder Magic or Thunder Rituals (*Lei-fa* 雷 法/*Wu-lei fa* 五 雷 法) constitute an established element in Taoist religious culture. Many ritual practices, convictions and emblems of Thunder Magic stem from antique traditions. The ritual specialisation and the name of Thunder Magic emerged during the Sung period (11th-13th cts.), uniting, developing and rationalizing exorcist and ritual methods that were already current in earlier periods of Taoist history albeit without the label of Thunder Magic.

This study intends to describe the basic notions, practices and intentions of Thunder Magic that often is connected with the names of *Shen-hsiao* (神 霄) and *Ch'ing-wei* (清 微) Taoism. The Taoist Canon contains abundant materials that show Thunder amulets, altars, seals and other ritual assets. Learned Discussions (*Lei-fa shuo* 雷 法 說) and similar texts show that Thunder Magic is not a unified religious and ritual element of Taoist culture but comprises ritual methods that many different local traditions, documentations and the patronage of legions of divinities identify.

Many texts cannot be attributed to any individual author but later Taoists, for example Pai Yü-ch'an (白 玉 蟬 fl. 1209-1224), are widely assumed to have had a hand in the actual formation of those texts. However, some texts very likely emerged in the time of Wang Wen-ch'ing (王 文 卿 1093-1153) who was a persuasive and famous promoter of Thunder Magic at the imperial court of the Sung (宋) dynasty. I try to focus on materials of that early period of Thunder Magic (12th ct.). Admittedly, the texts are often hard to understand and it is difficult to grasp the practical implications that they only indicate. In a sense, this presentation has a tentative character.

The German Research Foundation (Bonn) and the Ministry of Education of Taiwan (Taipeh) supported me to spend some time in Taiwan where I could consult Taiwanese colleagues and ritual specialists, who were very kind and helpful. In this sense, I especially wish to thank Profs. Lee Fong-mao (Academia Sinica) and Hsieh Ts'ung-hui (Taipei Normal University). The Katholische Akademische Ausländer Dienst (Bonn) gave Prof. Li Yüan-kuo (China/Chengdu, Academy of Social Sciences) the financial support to spend a couple of weeks in Berlin and work with me on some of the Thunder Magic materials that I present in this book. Yet, this book is just an attempt on my part to sort out the basic conditions of Taoist Thunder Magic. I hope to draw the attention of the scholarly world to this fascinating field of research that matters for the understanding of Taoist religious culture as we have it today. Concerning the publication of this book I wish to express my sincere thanks for the support by Harrassowitz Company (Wiesbaden), especially Mr. J. Fetkenheuer.

Berlin 2007   Florian C. Reiter

Introduction

Throughout history, the workings of religious Taoism are present on all levels of the Chinese society. Canonical hagiographies of saints and immortals and the biographies of historic persons show the activities and subtle influences that Taoists exerted. We usually read that they were learned persons who mastered the scriptures of Taoism and Confucianism, and they used to employ a great variety of practical skills and rituals. Skills and rituals saved from epidemics, droughts, floods and illness. Taoists acted for the wellbeing of the people, the administration and the state at large. This seems to be the standard range of Taoist aspirations and activities, and it is for this reason that we easily find the appropriate and persuasive rhetoric in Taoist texts throughout all ages.

Never mind the time and the period, Taoists always were very expert at divination. They healed illness and exorcised baleful influences, using amulets and other ritual devices. The Taoist activists seem to make use of a rather secret personal way of life, of prayer and meditation. There is, of course, the need for a protracted and rather complicated education that may lead to the career as a priest (*tao-shih* 道 士), and again there is some secrecy about the actual reality of this career.[1] However, we also notice that in many cases we never learn whether the respective Taoist was a priest or not, whether he knew how to perform rituals or not. Perhaps there was the general understanding that this profession had to be taken for granted.

Some canonical encyclopaedias present superb literary summaries of the Taoist religious culture. Sometimes they were compiled at the behest of emperors and rulers. The series of such encyclopaedias started with the title *Wu-shang pi-yao* (無 上 必 要6ct.). Encyclopaedias turn out to be major stepping-stones when we take up our studies of religious Taoism.[2]

On the other hand, biographies and hagiographies often speak about alleged historical persons and pose to feature the actual combination of all those

---

1   F.C.Reiter: The Aspirations and Standards of Taoist Priests in the Early T'ang Period, pp.133-150, in: AAS 1.
2   J.Lagerwey : *Wu-shang pi-yao*, somme taoiste du VIe siècle. Paris 1981.

theoretical and practical elements that characterize the Taoist and his vocation. Such sources show that the Taoist can interfere in the course of nature, for example, when he organises his individual essences of life, which goes with the Taoist self-cultivation to strive after an individual immortality. Taoism is characterized by elusive concepts of immortality that point to the very individual and final goal.

The Taoist, who lives in rural settings or even at the imperial court, is employed to fight against rain, drought and other natural disasters that need the deployment of martial spirit forces. Biographies and hagiographies bring us to realize that there must be a special relationship between the individual quality of the practitioner and the outer or social realities of his Taoist activities. We can find elaborate explanations for this delicate relationship in various canonical sources, and some of them are quite prominent in a book that intends to describe the conditions of Thunder Magic or Thunder Rituals.

Many activities in Thunder Magic remind of ecstatic and exorcist performances that certainly call to mind the opaque sphere of shaman culture. Taoist sources on the other hand explicitly reject the idea that Taoists have anything to do with the shaman branches of religious proficiency. We may suppose that the ecstatic state of mind and exorcist performances can hardly be a concern for an educated and learned person. Taoism, however, proves such a supposition to be wrong.

This study of Basic Conditions of Taoist Thunder Magic elucidates the workings of Thunder Magic (*wu-lei fa* 五 雷 法). This branch of ritual proficiency is associated with other schools in Taoist history, namely the *Shen-hsiao tao* (神 霄 道) and *T'ien-hsin cheng-fa* (天 心 正 法).

Since the Sung-period the name Thunder Magic is a general and comprehensive name for a vast variety of exorcist rituals that later continued to flourish and were labelled *Ch'ing-wei* (清 微) school (14th ct.). Finally, Thunder Magic became part of the general Taoist religious culture without being explicitly named Thunder Magic and singled out for separate practices. We can study this development, for example, in present day Taiwan, which, however, does not exclude the conscious continuation of Thunder Magic by individual priests or specialists who may claim to have family traditions reaching back to Taoists of the Sung period. The translation "magic" for the Chinese word *fa* (法) is the formulation that I prefer. In fact, we deal with rituals. Thunder Rituals would also be a good name translating *lei-fa* (雷 法).

Introduction 3

We know that the emergence of Thunder Magic in the Northern Sung-period is based on antique traditions of interpreting and handling natural phenomena, which always was and is a major concern in Taoism. There were specific rituals that focussed on the destructive force of thunder and lightning. They were employed in remote periods of Chinese history. For example, the scholar Wang Ch´ung (王 充) in the Han-period disputes in his *Lun-heng* (論 衡) the existence of a divinity called Thunder Lord (*lei-kung* 雷 公), saying that thunders are simply fire (*huo* 火).[3] The statement by Wang Ch´ung (王 充) seems to prove that ages ago a spirit or divine Thunder Lord was an object for scholarly considerations and, most certainly, for religious veneration. The darkness of history conceals those antique rituals. Anyway, we know for sure that the actual tasks, the purposes and means of Thunder Magic were matters of great concern, and they were integral parts of Taoism long before Thunder Magic explicitly appeared.

The T´ang Taoist Chang Wan-fu (張 萬 福 fl.711-713) tells us: "concerning the comprehensive rule over all divinities and demons in heaven and on earth, concerning the killing and subduing of wicked demons (*hsieh-mo* 邪 魔), the beheading and the annihilation of bad spirits (*yao-ching* 妖 精), the recruitment of divine forces (*ling* 靈) and the summoning of vital forces (*ch´i* 氣), the realisation of control and order for mountains and rivers, the cleansing of filthy atmospheres, the dispatch of petitions and the employment of postal [spirit] agents, the direct communication with the immortals (*shen-hsien* 神 仙), there is just nothing that has any priority to [the methods and rituals of] Orthodoxy and Unity (*Cheng-i* 正 一)." Ages later in the Sung period, Thunder Magic did serve well all these tasks.[4] It is quite telling that the statements of Chang Wan-fu (張 萬 福) read like an early summary of the programmatic range of the much later Thunder rituals. We want to keep in mind that there are also some other, additional aspects of the ritual practice in Thunder Magic that classify such rituals as proper parts of *Cheng-i* (正一) Taoism.

We consider that the Thunder specialists employed mostly secret skills and crafts that go with the name of Thunder Magic to solve individual and communal problems that people believed to have a spiritual or transcendent

---

3 Wang Ch´ung: *Lun-heng* 6, 23, 96-103, esp. pp. 100-101 (*Lei-hsü p´ien*). Shanghai 1974.
4 TT 1241 *Ch´uan-shou san-tung ching-chieh fa-lu lüeh-shuo* 1.4a-4b, following the text *Cheng-i fa-wen k´o-chieh p´in* (正 一 法 文 科 戒 品). Concerning this text see, U.A.Cedzich, p. 458, in: K.Schipper and F.Verellen eds.: The Taoist Canon, A Historical Companion to the Daozang (henceforth *Companion*). 3 vols. Chicago 2004.

cause. For example, droughts and floods are such problems that can endanger the wellbeing of agrarian communities. The ritual help by means of Thunder Magic, however, is most seldom a communal event in which the local population can participate, in one way or other joining the festivities. Thunder Magic is a rather personal and secret matter. The communal participation is restricted to a few sections of the old and established thanksgiving rituals (*chiao* 醮). When the Thunder specialist and priest had secretly performed his Thunder rituals and had been successful, the community may then stage such communal rituals.

Before taking up any details, I present the biography of the Taoist Yeh Ch´ien-shao (葉千韶) who lived in the 9th century, well before the heyday of Thunder Magic. His biography substantiates most of the aspects of Taoist culture that I addressed so far. The text also unites many characteristics that generally describe the Taoist priest and exorcist.

The practical and ritual means that were at the avail of Yeh Ch´ien-shao (葉千韶) have a long history in China. They stem from historical periods before the reputed beginning of organized Taoism in the 2nd century A.D. The Sung (宋) period started in 960 A.D., about one hundred years after the life of Yeh Ch´ien-shao (葉千韶) who shows activities and spiritual potentials that forecast specific elements of the later Thunder Magic. His biography indicates the personal qualities that characterize the practitioners of Thunder Magic. The term Thunder Magic stands for the attempt to rationalize a welter of long established religious notions and rituals, which may be due to the intellectual and rationalistic disposition of the Sung (宋) period. We also remember that emperor Sung Hui-tsung (宋徽宗 r. 1100-1126) developed a great personal interest in the Taoist religion and actually thought himself and his staff to represent the heavenly spirit-administration. On the other side, the priest and Thunder specialist professes to embody a spirit-career with ranks and varied promotions, which enables him to live up to the extraordinary standard of being divine. Our sources do not let us have any doubts about this claim.[5] We notice that we find in *A Dictionary of Official Titles in Imperial China* many of the spirit-titles that refer to the assumed power structure of Taoist Thunder specialists and their deities.[6]

---

5   See the grand spirit-ranks and promotions for the Taoist priest in TT 1220 *Tao-fa hui-yüan* 56.39a-39b. For a translation of this chapter in TT 1220 see below, Chapter II: The Scope of Taoist Thunder Magic.
6   See Charles O. Hucker, Stanford 1985 (henceforth *Hucker*).

Introduction 5

The biography of Yeh Ch´ien-shao (葉 千 韶) serves us well as an entertaining example to illustrate the practical and visible role of Taoist Thunder Magic. The biography is contained in the collection "Supplementary Lives of Immortals" (*Hsü-hsien chuan* 續 仙 傳) by Shen Fen (沈 汾) of the Southern T´ang period (937-975). The biography of Yeh Ch´ien-shao (葉 千 韶) is one of the twelve items in chapter 2: "Hidden Transformations" (*yin-hua* 隱 化).

The Biography of Yeh Ch´ien-shao (葉 千 韶) [7]

(17a) "Yeh Ch´ien-shao (葉 千韶) had the name (*tzu* 字) Lu-ts´ung (魯 聰) and hailed from Chien-ch´ang (建 昌) district in Hung-chou (洪 州).[8] When he was young he adhered to the Taoist skills (*tao-shu* 道 術) of his [spiritual] teacher masters, the [two] Taoists from the Western Mountains (hsi-shan 西 山), the Perfect Lords (*chen-chün* 真 君) Hsü [Sun] (許 [遜]) and Wu [Meng] (吳 [猛]).[9] He abstained from cereals and practiced breathing techniques (*fu-ch´i* 服 氣).

---

7 See *Yeh Ch´ien-shao* in: TT 295 *Hsü-hsien chuan* 2,16b-18b. Also see *Hsü-hsien chuan* 2, 13a-13b; in: Hsiao T´ien-shih ed.: *Li-tai chen-hsien shih-chuan* (*Tao-tsang ching-hua* 5/7, Taipei 1980); Concerning TT 295, see F.Verellen, *Xu xian zhuan*, pp.429-430, in: *Companion*, he also indicates other editions of this title that are either complete or fragmentary. See *T´ai-p´ing kuang-chi* (太 平 廣 記) 394 (*lei* 雷 2), 6a-6b (p.1607, ed.: Kyoto 1972) contains the story of a certain Yeh Ch´ien-shao (葉 遷 韶) who was initiated by Lei-kung (雷 公) as to be able to summon thunders and to save people. The deity explained that he had five brothers. The two Yehs (葉) most likely are the same person, although their places of origin read different. Both names most certainly stand for the one person who hailes from Chiang-hsi province. The background of the person in TT 295 sounds better and more learned, whereas the *T´ai-p´ing kuang-chi* (太 平 廣 記) of Li Fang (李 昉, 978 CE) speaks about a young lad who collected firewood when he experienced that divine encounter. I translate the TT 295 version.
8 This is today´s Nan-ch´ang district in Chiang-hsi province, where Wang Wen-ch´ing (王 文 卿) also hails from.
9 Concerning the two famous Taoists (3rd ct.) who allegedly were experts in Thunder Magic, see TT 1220 *Tao-fa hui-yüan* 56.10a; for a complete translation of this chapter (nr.56) in *A Corpus of Taoist Ritual* see below Chapter II. See F.C.Reiter: „The Name of the Nameless and Thunder Magic", p.115 in: P.Andersen and F.C.Reiter eds.: Scriptures, Schools and Forms of Practice in Daoism, A Berlin Symposium, in: AAS 20. TT 1220: 125.1a sq. *Chiu-chou she-ling man-lei ta-fa* (九 州 社 令 蠻 雷 大 法) shows Hsü Sun (許 遜) to be the very first spiritual patron saint of the Earth Altar (Thunder) rituals. The phrasing in TT 295 is opaque. It literally says: "…he served the Taoists…as his teacher masters. His Taoist skills [were] the avoidance of cereals and the breathing technics …" Yeh Ch´ien-shao (葉千韶) reputedly lived in the T´ang-period, a few centuries after the times of Hsü and Wu. Concerning the Altar of Earth Thunder, see below Chapter II, rf. TT 1220:

Once he stayed alone in a mountain when suddenly storm and rain came up together with thunder and lightning. There was a person clad in white garments who [appeared and] reverently addressed Yeh Ch'ien-shao (葉千韶) saying that his Taoist virtuous ways were accomplished to the highest degree, and the registers of the immortals have recommended him for the ascent [to the immortals]. [Now,] in the world of man he should employ demons and deities as his servants and emissaries and so continue furthermore to display his meritorious deeds. Today, a divine person (shen-jen 神人) is about to descend, and he, Yeh Ch'ien-shao (葉千韶), would be able to meet the divine person. He should not be afraid at all.

Yeh Ch'ien-shao (葉千韶) thereupon burnt incense and retired with folded hands to practice silent meditation. In a moment, a perfected [spirit] official who was clad in red garments (chu-i chen-kuan 朱衣真官) descended, arriving from the clouds far away. More than ten [spirit] generals additionally [descended] carrying swords, and on their belts they all carried the dragon-and-tiger amulet (lung-hu fu 龍虎符). The accompanying units (pu-ts'ung 部從) of demoniac and divine [troops] were legions. There were also two attendants who were clad in yellow and green garments, and each of them held one role of registers (pu i chüan 簿一卷) [from the heavenly archives]. The divine generals (shen-chiang 神將) lined up to salute Yeh Ch'ien-shao (葉千韶). The perfected [spirit] official [with red garments] addressed him and said:

"[Obeying] the decree of heaven we transfer to you these registers (pu 簿). [We], the divine generals, emissaries and troops are obliged to provide perfectly any service as your emissaries in order to save your contemporaries."

(17b) Yeh Ch'ien-shao (葉千韶) reverently received the heavenly documents (t'ien-shu 天書), held them in both hands and perused them. They resembled worldly military registers (ping-chi 兵籍). The attendants who firmly held the registers and documents asked [Yeh Ch'ien-shao (葉千韶)] for his summons, so that they could execute his orders.

After [the encounter] Yeh Ch'ien-shao (葉千韶) could shout long screams,[10] and in this way, the wind was aroused in forests and abysses. When he spit out

---

56.13a.
10 Or translate: „he roared". Shouts or the loud, almost explosive dispatches of concentrated breaths are a rather usual feature of some Taoist techniques.

water,[11] rain fell in the plains. When he heavily pressed down his feet onto the ground, the noise of thunder evolved [resembling the noise of a] windlass. When [he seemed to] paint with his hand in empty space,[12] radiant lightning would occur which baffled everybody.

Yeh Ch´ien-shao (葉千韶) roamed then through the world and feigned to be mad. Often being drunk, he popped around in thoroughfares. When he suddenly made loud and threatening noises, he seemed to shake with might. When he was questioned what it was all about, he gave this sort of answer: "I saw in this or that place a fire", or "[I saw] in this or that place a drought and I [just] dispatched rain to save [the situation]". When the people set out to search for the facts, they always found [his words] to have been true.

Occasionally he passed some prefectures and districts and they all asked him for help when they were suffering from a drought. Yeh Ch´ien-shao (葉千韶) would prepare an altar (*hsiang-an* 香案) and start to speak spells (*ch´i-chou* 啟咒). Rain would begin to fall within a short moment. When people asked him to let thunders [rumble], he would press his feet onto the ground and consequently the sound [of thunders] would emerge from below in the earth like [the rolling of] a windlass.

In other cases when bad rainstorms had occurred and the prayers for a clear sky got no response at all, [the people] would ask Yeh Ch´ien-shao (葉千韶) to stop it. Consequently he would perform his [exorcist] methods (*tso shu* 作術) **(18a)** and achieved that the sky cleared up (*ch´ing-chi* 晴霽). When a drought occurred during the wintertime and prayers for snow were to be offered Yeh Ch´ien-shao (葉千韶) wore simple unlined clothes and stood barefooted in the sun. He would scream and recite [spells], and within a moment, wind and clouds would assemble and snow would fall throughout all the night.

Furthermore, he used amulets (*fu* 符) to save [and heal] illness and distress. He did not wait for people to come up to him and ask for help, but when he saw a sick person, he always felt sympathy and saved then that person. When wicked *mei*-demons (*hsieh-mei* 邪魅) were around and heard the name of Yeh Ch´ien-shao (葉千韶), they spontaneously reformed their ways to the better. *Mei*-

---

11 Possibly this means amulet water.
12 This means the drawing of physically unseen amulets in empty space.

demons (*hsieh-mei* 邪 魅) that were treated with [his] amulets did not again display [any wicked might] until the end of their days.[13]

In the eleventh year of the reign title *hsien-t´ung* (咸 通 11; 870 A.D.) he set out to travel and finally reached Hao-chou (濠 州),[14] where he learnt that the governor Liu Fang (劉 昉) suddenly had been struck by a wind and was about to die. Some famous physicians had already been called in but they could not cure the governor.

Yeh Ch´ien-shao (葉 千 韶) entered with his travel stick [Hao-] chou (濠 州) and said [to Liu Fang (劉 昉)]: "[You] induced me to come here in order to make you come back again to life". Thereupon he wrote three amulets that he fastened on the top of the shoulder [of the governor], on his ribs and on a leg and said: "I force the wind to come forth from your feet. In three days [you] should be well again." Finally, the wind made a whizzing sound and came forth from the hollows of the feet of the governor. [The governor] recuperated after three days and was well as before.

Liu Fang (劉 昉) was a very learned person. He was well familiar with literature and was generally dedicated to Taoist crafts (*tao-shu* 道 術). He made an official career and attained the prefecture [as governor] where he exerted a good government that reached the common folks. He once said to his guests and attendants **(18b)**:"in all my life, I took the *Tao* to be my teacher. Once a violent wind suddenly hit me and thereupon I experienced the impact of a saint (*sheng-jen* 聖 人) who made use of some emblems (*hsiang* 相) to save and heal me.[15] [The experience] that [a man like] Tung Feng (董 奉) returns harmonised *hun*-souls (魂) to a man (*shih* 士), this really is something one has to wait for.[16] In fact, this was a response due to the strength of *Tao* (*dao-li* 道 力)". The people in the prefecture sensed that divine qualities were at hand in the person of Yeh Ch´ien-shao (葉 千 韶). Liu Fang (劉 昉) welcomed then Yeh Ch´ien-shao (葉

---

13 This speaks about demons in the shape and with the face of humans.
14 This is a location in An-hui province.
15 This (*hsiang* 相) should refer to the amulets.
16 Here, Tung Feng (董 奉) most certainly is a personal name that points to a Taoist healer who was a rather famous figure in Chiang-hsi province. He is said to have brought back to life some other high official, see TT 1248 *San-tung ch´ün-hsien lu* 4.5a-5b; TT 1139 *San-tung chu-nang* 1.18b-19a. Especially see, Ch´en Shu-yü (陳 舜 俞): *Lu-shan chi* (廬 山 記) 2.2a-2b, in: *Pi-chi hsü-pien*. Kuang-wen Comp. Taipei 1969. See F.C.Reiter: Der "Bericht über den Berg Lu (*Lu-shan chi*) von Ch´en Shun-yü, ein historiographischer Beitrag aus der Sung Zeit zum Kulturraum des Lu Shan, pp.117-119. München 1978.

千韶) at a fasting festivity (*chai* 齋) in the prefecture and expressed his wish to serve him as his teacher master. He gave him a lot of gold and brocade to express his thanks. Yeh Ch'ien-shao (葉 千 韶), however, speedily deserted Liu Fang (劉 昉) and disappeared. A search for him was made but no traces [of the Taoist] could be found.

Later, in the regions of Ching and Hsiang (荊 襄) some persons saw [him] when he talked laughing about the events in Hao-chou (豪 州). It was evident that he was hiding away in the Western Mountains (hsi-shan 西 山) for more than ten years. Among the people of today there are some persons who had eventually seen him". ***

Yeh Ch'ien-shao (葉 千韶) relies on the spiritual and perhaps even practical guidance of two Taoists of old, namely Wu Meng (吳 猛) and Hsü Sun (許 遜). He finds his motivation in their reputed successes, practices the abstention from cereals and breathing techniques (*fu-ch'i* 服 氣).[17] Yeh Ch'ien-shao (葉 千 韶) must have been an advanced Taoist, which the apparition of a company of spirit administrators and guards seems to prove. They appeared as a divine response in honour of the Taoist success of Yeh Ch'ien-shao (葉 千 韶). Perfected spirit officials with an entourage of warriors and emissaries entrusted him with the registers of divine martial forces that from now on would obey his commands. These spirit forces most likely were Thunder deities, because the Taoist started

---

17 This is *pars pro toto* for his self-cultivation. Wu Meng (吳 猛) hailed from Yü-chang (豫 章). He was an exorcist, who could subdue epidemics and had access to the world of the immortals, see Ch'en Kuo-fu: *Tao-tsang yüan-liu k'ao*, pp.461-462 (*Kao-hsien chuan*). Rpr.Taipei 1975. See F.C.Reiter: "Die Ausführungen Li Tao-yüans zur Geschichte und Geographie des Berges Lu (Chiang-hsi) im „Kommentar zum Wasserklassiker", und ihre Bedeutung für die regionale Geschichtsschreibung", p. 20, in: *Oriens Extremus* 28/1 (1981). The canonical biography of the temporary Magistrate of Hsi-an, Wu Meng (西 安 令 吳 猛), is TT 296 *Li-shih chen-hsien t'i-tao t'ung-chien* 27,1a-2b. At the age of fourty he received the "divine reciepies" (*shen-fang* 神 方) of the perfected Ting I (丁 義). He took as his teacher master the governor of Nan-hai (南 海 太 守) Pao Ching (鮑 晴), who passed on to him some unspecified "secret methods". During the period Huang-lung (吳 黃 龍, 229-232) "heaven revealed the White Cloud Amulets (*pai-yün fu* 白 雲 符)" that enabled him "to practice greatly his Taoist skills (*tao-shu* 道 術)". Later (Chin 晉) "the Perfected Lord Hsü (許 真 君) joined him and received then his complete Taoist professional secrets (pi-yao 秘 要)". The biography reports his magic and exorcist abilities. For example, Wu Meng (吳 猛) stopped stormy winds with the application of his amulets. The geographic indication Western Mountains (西 山) may allude to the mounatains "west of the great Chiang", in this case the mountains west of Nan-ch'ang might be the actual point of reference (南 昌 西 山). Also, see, J.M.Boltz: A Survey of Taoist Literature, Tenth to Seventeenth Centuries, pp. 70-72, concerning the two Taoists. Berkeley 1987.

to achieve successes that in later times would have been identified with Thunder Magic. He also acquired a new personal quality and rank, which he outwardly wished to conceal and therefore he behaved like a lunatic. He could enter the mental and physical state of ecstasy, screaming, spitting and stampeding with his feet. In this way, he could procure magic responses in the cosmos. He could see what was going on in remote places far away, and so he dispatched rain, for example, to relieve a drought or fires in those places. Rain, fair weather, snow in winter, anything what was needed Yeh Ch´ien-shao (葉 千 韶) knew how to handle the situation. He may show up barefooted and clad in ragged clothes, possibly with dishevelled hair, and perform rituals to improve the weather conditions.

In any case, he performed specific skills (*tso-shu* 作 術) to achieve the desired results. The writing of amulets was one of his crafts. He used them to subdue demons in human shape and appearance. It is significant that his amulets were used to heal illness. The Taoist would offer such help without being asked for, and he was even mighty enough to save a person from the danger of imminent death.

Anyway, Yeh Ch´ien-shao (葉 千 韶) was not a spectacular Taoist who would have enjoyed the attention of the social elite and the court. He rejected profane goods in return for spiritual help and preferred seclusion when he felt uncomfortable about public attention.

The biography mentions *en passant* an official Taoist fasting and purification *chai*-liturgy (齋) at which the Taoist participates. All his other activities can be labelled magic or exorcist, and in fact, exactly this feature is very much in the focus of the biography. The text does not give any explanation or assessments. We observe that his magic and exorcist activities go with the installation of altars and offerings of incense. They are clearly structured religious acts. I find it most important that the exorcist actions of Yeh Ch´ien-shao (葉 千 韶) are based on heavenly documents (*t´ien-shu* 天 書), which are the registers of martial spirit forces. It is also a characteristic feature of the historically later Taoist Thunder Magic to have such registers (*lu* 錄). We notice that descriptions of such ritual elements are contained in many biographies in the Taoist Canon. They are the standard assets of Taoist professional life. It is very much rewarding to study the canonical explanations and descriptions of the spiritual forces and the suitable ritual means, which characterize Taoist religious culture.

Chapter I: Man and Thunder Magic

A first approach to Thunder Magic

Thunder Magic is an elusive term that covers a huge array of Taoist ritual practices. Most of them seem to serve exorcist purposes, and therefore they may not enjoy the broad scholarly attention that the standard repertoire of Heavenly Master Taoism receives.[1] Hagiographical and biographical sources show that the Taoist participation in social life implies much more than the mastery of the liturgy of great festivities that the terms *chiao* (醮) and *chai* (齋) identify.[2] Taoist priests traditionally operated as faith healers. They also were keen to lend their helping hands when droughts, floods and other unpleasant or disastrous events marred daily life. Taoists often seem to have used exorcist rituals to expulse malignant influences and their causes that people believed to be the influences of wicked demons.[3]

This situation is well documented in the history of the Taoist religion. I list just a few renowned sources that provide such information. There is, for example, the Scripture of the Great Peace (*T'ai-p'ing ching* 太平經) that is connected with the Taoist history in the second century of the Christian era.[4] The later encyclopaedias *San-tung chu-nang* (三洞珠囊), *Tao-tien lun* (道典論) and *Tao-chiao i-shu* (道教義樞)[5] list numerous entries that tell us how Taoists dealt

---

[1] We have some remarkable Chinese studies, for example, the two monographs: Li Yüan-kuo (李遠國): *Shen-hsiao lei-fa* (神霄雷法). Chengdu 2003; and Liu Chung-yü (劉仲宇): *Tao-chiao fa-shu* (道教法術). Shanghai 2002. Also see L.Skar: "Administering Thunder: A thirteenth Century Memorial Deliberating the Thunder Rites", in: *Cahiers d'Extrême-Asie* 9, pp.159-202 (1996-1997). And the same "Ritual Movements, Deity Cults, and the Transformation of Daoism in Song and Yuan Times", pp.413-463, in L.Kohn ed.: Daoism Handbook (Leiden 2000). A rough survey on Thunder Magic materials and related sources in the Taoist Canon gives K.Schipper and F.Verellen eds.: *Companion*, pp.1384-1386.
[2] See M.R.Saso: Taoism and the Rite of Cosmic Renewal. Washington 1972. Also see K.Schipper: Le corps taoiste. Paris 1982, J.Lagerwey: Taoist Ritual in Chinese Society and History. New York 1987, R.Hymes: Way and Byway, Taoism, Local Religion, and Models of Divinity in Sung and Modern China. Berkeley 2002; and F.C.Reiter: Religionen in China, Geschichte, Alltag, Kultur. München 2002.
[3] F.C.Reiter: Der Perlenbeutel aus den Drei Höhlen (*San-tung chu-nang*), Arbeitsmaterialien zum Taoismus der frühen T'ang Zeit, p.9 sq. („Hilfe und Führung"). AF 112.
[4] K.Schipper: "Taiping jing", pp.277-280, in: *Companion*. See B.Hendrischke: The Scripture on Great Peace, The *Taiping Jing* and the Beginnings of Daoism. Berkeley 2006.
[5] See note 3, and see F. C. Reiter, pp.440-441; H.-H. Schmidt, pp.445-446, 442, in: *Companion*. See F.C. Reiter: Kategorien und Realien im Shang-ch'ing Taoismus (*Shang-ch'ing tao lei-shih hsiang*), Arbeitsmaterialien zum Taoismus der frühen T'ang Zeit. in: AF 119; Wang Zongyu (王宗昱): *Daojiao yishu yanjiu* (道教義樞研究). Shanghai 2001.

Chapter I: Man and Thunder Magic

with the spirit world, mostly quoting scriptures dating from the *Nan-pei ch'ao* (南北朝) period.

In the Northern Sung Period during the reign of emperor Sung Hui-tsung (宋徽宗 rg.1100-1126) a welter of old concepts and practices came to be assembled under the name of Thunder Magic (*wu-lei fa* 五雷法). Wang Wen-ch'ing (王文卿 1093-1153) effectively managed to perform Thunder rites and enthralled the emperor with his obvious successes that were believed to result from his ritual expertise. The Taoist eliminated fox spirits and at one time enforced good weather conditions for state rituals. Wang Wen-ch'ing's (王文卿) ritual efforts are documented in his canonical biography and in *A Corpus of Taoist Ritual* (*Tao-fa hui-yüan* 道法會元) of the 14th century. [6]

This literary collection contains quite a number of texts from the School of Pure Subtlety (*Ch'ing-wei* 清微) that from the 13th century onwards continued and extended the traditions of Sung Thunder Magic. It is well known that a remarkable number of canonical texts are either directly connected with Thunder Magic or can be associated with the concepts of Thunder Magic. [7] Very long periods of Taoist history and activities elapsed until the 12th century when literary and rather comprehensive efforts were made to assemble and formulate what could be said about Thunder Magic. Many of these materials are hard to date, even if we can place them within the frame of about two hundred years, which means from the 12th to the 14th centuries. The exact date of origin of quite a few tracts on Thunder Magic in *A Corpus of Taoist Ritual* remains uncertain.

We have some texts in *A Corpus of Taoist Ritual* that explicitly indicate Wang Wen-ch'ing (王文卿) to be the author, and I accept this information as long as textual evidence does not contradict the attribution. Studying these texts, I

---

6   K.Schipper and Yuan Bingling, pp.1105-1113, in: *Companion*. P. van der Loon: "A Taoist Collection of the Fourteenth Century", pp. 401-405, in: W.Bauer ed.: Studia Sino-Mongolica, MOS 25. F.C.Reiter: "A Preliminary Study of the Taoist Wang Wen-ch'ing (1093-1153) and his Thunder Magic (*lei-fa*)", in: *Zeitschrift der Deutschen Morgenländischen Gesellschaft* (henceforth ZDMG), 152, pp.155-184. And the same, "The Discourse on the Thunders 雷說, by the Taoist Wang Wen-ch'ing 王文卿 (1093-1153)", in: *Journal of the Royal Asiatic Society* (henceforth JRAS), 14/3, pp.207-229.

7   Compare *Companion* pp.1384-1385. See F.C.Reiter:: Grundelemente und Tendenzen des religiösen Taoismus, das Spannungsverhältnis von Integration und Individualität in seiner Geschichte zur Chin-, Yüan- und frühen Ming-Zeit, pp.42-55 (concerning Qingwei 清微), in: MOS 48.

describe some basic features and specific activities. They all are closely interwoven and reveal the conditions and workings of Thunder Magic.

The Taoist patriarch Chang Yü-ch´u (張 宇 初 1361-1410) acted on imperial orders in 1406 when he started to recompile the Taoist Canon after its destruction under Mongol rule in the 13th century. He classified Thunder Magic rituals as minor rituals (*hsiao-fa* 小 法), which seems to be a belittling qualification. This assessment does in fact not say anything about the religious and spiritual standards of such minor rituals which I have shown somewhere else.[8] In the book Basic Conditions of Taoist Thunder Magic, I present texts that reveal the religious and spiritual quality of Thunder Magic. I consider various practical aspects, completely basing myself on canonical sources that we can associate with or even attribute to Wang Wen-ch´ing (王 文 卿). I present information to show that such minor rituals actually comprise the whole range of well-established Taoist notions and practices. They are demanding as far as the religious and personal standing of the performing priest is concerned, which again Wang Wen-ch´ing (王 文 卿) perfectly exemplified.

Human Nature and Thunder Magic

Wang Wen-ch´ing (王 文 卿) presents a fair number of theoretic expositions that elucidate basic and ritual actions of Thunder Magic. Many of these texts are didactic expositions and pointedly explain rather specific themes. The answers that Wang Wen-ch´ing (王 文 卿) gives his students are to the point, and yet, they should not be understood as the final solutions to the questions that originally were raised. We certainly miss the oral and secret instructions of the Taoist that he may have passed on to his advanced disciples. However, his answers are always conclusive and therefore we have a good chance to collect a lot of information that should help us understand the basics of Thunder Magic. I start out enquiring about the practical concerns and activities of a Thunder specialist.

The Thunder specialist is a Taoist priest (*tao-shih* 道 士) who performs exorcist crafts. He summons individual divine forces in order to employ them for ritual

---

[8] F.C.Reiter: "The Management of Nature: Convictions and Means in Daoist Thunder Magic (Daojiao leifa)", pp.193-210, in F.C.Reiter ed.: Purposes, Means and Convictions in Daoism, A Berlin Symposium, in: AAS 29; and see my Grundelemente und Tendenzen des religiösen Taoismus, pp.11-35; 132-139, in MOS 48.

purposes. He first addresses divine and transcendent entities that are not at all beyond the confines of his own human body. We know that the Taoists developed and fostered specific convictions concerning the divine nature of the human being, or to be more precise, of "the self" (*tzu-chi* 自 己). This conviction is the absolute basis for any ritual activity, which is easy to illustrate. We find, for example, in the following short essay the explicit identification of the individual self of the priest with the Divine Norm (*ling-far* 靈 法) of Thunder Magic.[9]

The Secret Instructions Concerning the Thunder Rituals (*lei-far pi-chi* 雷 法 秘 旨) (2b) explain: the self is the Divine Norm (*ling-far* 靈 法). As to the divine self (*ling-woo* 靈 我), if it is not divine, what [else then] could unite with the thunder divinities (*lei-sheen* 雷 神)? I (*wo* 我) am capable to excite [them] in a good way, and the thunder divinities respond in a good way. One appeal to [the thunder divinities] has one response, and thus absolutely everything is settled. When there is a [right] mind [at work], it can appeal to the divinities. When the divinities turn away and do not response, the appeal was done without a proper mind. The response [of the divinities] is like an echo, and just for nothing, there is also no response at all. You must not have any erratic thoughts. The completely true mind (*i-p'ien chen-hsin* 一 片 真 心) must be completely free.[10] [In this way] the mind and the thunder divinities intermingle and are just like one united entity. In this way, I am the thunder divinity, and the thunder divinity is I. All the [divine] responses are concurrent with me. There is no response that would not occur, (3a) [for example], sympathy, compassion, profit and help. Above, I unite with the heart of heaven (*t'ien-hsin* 天 心).[11] My love for the living beings is my virtue, and it is for this reason that heaven certainly does not disregard me. Thunders receive respectfully the heavenly orders. How could they disregard me?[12]

---

9 TT 1220: 77.2b-3a. See TT 1220: 84.8b, referring to the encounter of Wang Wen-ch'ing (王 文 卿) with his teacher master (see below) and the transmission of the title (*lei-fa pi-chih* 雷 法 秘 旨). However, we cannot prove the identity of the extant texts in consideration and the allegedly transmitted text.
10 "free" or "unknowing".
11 Compare TT 1220: 56.4b, see below the translation of *chapter 56*. For this term see, P.Andersen: "The Practice of *Bugang*", in *Cahiers d'Extrême-Asie* 5, p.28 (1989-1990).
12 The text *Tao-miao* (道 妙; TT 1220: 84.2a-5b) extends this notion by identifying the "self" (*wo* 我) with the ancestral breaths (ch'i che sheng-i chih tsu 氣 者 生 一 之) and also with heaven (t'ien 天) implying all its potentials (p.3b). It is clear that the tract is written ages after the time of Wang Wen-ch'ing (王 文 卿). Also compare TT 1015 *Chin-so liu-chu yin* 5.1a-1b (t'ai-hsüan yüan-ch'i suo-sheng san-yüan yin 太 玄 元 氣 所 生 三 元 引), with a definition of the spiritual force (shen 神), which surfaces in the practices and convictions of Thunder Magic. For this text is attributed to Li Ch'un-feng (李 淳 風, 602-670), see

The Secret Instructions Concerning the Thunder Rituals 17

The subsequent explanation concerning this text says that the Fire Master, the perfect [teacher master] Wang (汪),[13] personally transmitted it to Shih-ch´en, Wang Wen-ching (侍 宸, 王 文 卿). This statement intends to show that the text itself was a secret transmission in the school of Thunder Magic (*lei-t´ing chih pi-ch´uan* 雷 霆 之 秘 傳).

The text Secret Instructions Concerning the Thunder Rituals (*lei-fa pi-chih* 雷 法 秘 旨) brings us to consider the very nature of the relationship between the acting priest and the cosmic or absolute divine being that adopts the apparitions and potentials of Thunder deities. "The [Thunder] emissary, this is just me. I am the one who is the emissary and I shall act it out until the end of my life", says the theoretical tract Mysteries of Tao (*tao-miao* 道 妙).[14] Obviously, there is an osmotic relationship or even a special unity between man and the divine that Thunder deities identify.

We generally observe the Taoist inclination to think in terms of a grand unity, which surfaces in the chronologies of Taoist schools and affiliations. I remind of the hagiography of the Ch´ing-wei (清 微) School (*Ch´ing-wei hsien-p´u* 清 微 仙 譜) or the hagiographies of the Ch´üan-chen (全 真) School (*Chin-lien cheng-tsung-chi* 金 蓮 正 宗 記 / *Chin-lien cheng-tsung hsien-yüan hsiang-chuan* 金 蓮 正 宗 仙 源 像 傳).[15] The hagiographies of T´ai-shang Lao-chün (太 上 老 君) show this divinity to be the unifying body of the universe and the moving power of creation. T´ai-shang Lao-chün (太 上 老 君) is the leading expression of that general Taoist notion.[16] We certainly understand that scriptures and rituals

---

P.Andersen, in: Companion pp.1076- 1079.
13 TT 1220: 76.3a indicates the taboo name Tzu-hua (子 華), the tzu-name Shih-mei (時 美). See TT 1220: 83.1a, which is a later tradition. It says that the taboo name is Shou-chen (守 真) "Guarding Perfection").
14 TT 1220: 84.4b-5a., and see below.
15 Concerning TT 171; 173; 174, see F.C.Reiter, pp. 1100, 1135-1136, 1136-1137, in: *Companion*. Also see F.C.Reiter: Grundelemente und Tendenzen des religiösen Taoismus, pp.42-76.
16 See A.K.Seidel: La divinisation de Lao Tseu dans le taoisme des Han, p. 84 sq., 92 sq. (Paris 1969). F.C.Reiter trl. ed.: Leben und Wirken Lao-Tzu´s in Schrift und Bild (*Lao-chün pa-shih-i hua t´u-shuo*), pp.21-239 (Würzburg 1990). Also see, for example, TT 1220: 84.6a, explaining that the three "venerable divinities" of the Three Caves (*san-tung* 三 洞) are to be identified with primordial forms of breath and the same time with the individual names of Thunder deities. "In fact, the three deities are in basically one deity, and the one entity is three deities" (*san-shen pen i-shen i-t´i san-shen shih yeh* 三 神 本 一 神 一 體 三 神 是 也). The statement also ascertains that Thunder Magic is a firm part of the established Taoism of the Heavenly Master (*cheng-i tao* 正 一 道 / *t´ien-shih tao* 天 師 道).

mirror this religious conviction of a cosmic unity of man and the divine, which Thunder Magic rituals set to use.

When we focus our study on Thunder Magic we find several tracts by Wang Wen-ch'ing (王文卿) that explain selected constituents of Thunder rituals. He also makes some statements about the spiritual conditions of the human nature that matter for any ritual performance. We come to understand the actual scope of the ritual profession in terms of intellectual and practical efforts.

Our first and most basic question that we formulate may refer to the individual standard and the quality of the Taoist. How does he summon and dispatch the divine forces that exist within his own human body and let them serve ritual purposes?

This sort of question came to the mind of the disciples of Wang Wen-ch'ing (王文卿) who worked out appropriate texts to feature the conditions of Thunder Magic and in this way presented conclusive answers. We are fortunate to have the didactic collection *Wang Shih-ch'en ch'i-tao pa-tuan chin* (王侍宸祈禱八段錦). His otherwise unknown disciple Yüan Wu-chieh (袁無介) formulates a number of interesting questions. For example, the paragraph "Summoning and Uniting" begins with a question that Yüan Wu-chieh (袁無介) sets forth. Wang Wen-ch'ing (王文卿) gives his explanation, and another unknown person with the name Yü-feng (御風) contributes additional commentaries.

Summoning and Uniting [the Thunder Divinities]
(chao-ho 召 合) [17]

Yüan Wu-chieh (袁 無 介) sets forth his question about the so-called summoning and uniting. He wishes to learn about the essential points of summoning and uniting.

[Wang] Shih-ch'en ([王] 侍 宸) says: The people of today, the gentlemen who practice Taoism only know that the general of the heart (hsin-chiang 心 將) comes out of the heart, the general of the liver (kan-chiang 肝 將) comes out of the liver and the general of the kidneys (shen-chiang 腎 將) comes out of the kidneys. Any summoning and uniting that is done in this way, is not in accord with the right principles (cheng-li 正 理). How does it come that people do not know that metal, wood, water, fire and earth, the generals of the five intestines all emerge from earth (t'u 土)? Earth is the leading force (chu 主) of the five elements (wu-hsing 五 行), and the spleen (p'i 脾) belongs [to earth]. [18]

The commentary by Yü-feng (御 風) says: the theory of summoning and uniting relies on the spells (chou-yü 咒 語) in rituals that are used to call out the summons. At the time of a union [with the divine], you first summon [the divinity]. When [the divinity] has arrived you inhale as to lead [the divinity] into the central palace (chung-kung 中 宮) to realize the union.

As a case in point, **(5b)** Marshal Teng (鄧 帥) belongs to [the element] fire and returns to the heart (hsin 心).

---

17   TT 1220: 69.5a-11a. Compare TT 1220: 124.2b-3a *Chao-ho pi-fa* (召 合秘 法), see F.C.Reiter: "A Preliminary Study of the Taoist Wang Wen-ch´ing (1093-1153) and his Thunder Magic (*lei-fa*)". In: ZDMG 152, p.172-173, note 54; p.179, note 82, rf. TT 1220: 85.7a-7b. The gist of this paragraph has been transmitted in later traditions, see TT 1220: 83.3a-3b *Chao-he hsüan-chi* (召 合 玄 機). This text features the self-identification with the divinity T´ien-mu (天 母). See TT 1220: 85.14a-18b (T´ien-mu 天 母/Pei-tou chih mu 北 斗 之 母). Ch. 83.2a-3a gives a grand description of the divine signs of the Heavenly Mother Mo-li-chih-t´ien ta-sheng (摩 利 支 天 大 聖) who is Mârici, a deity that unites Brahmanic mythology with Chinese Buddhist and Taoist notions, see W.E.Soothill and L. Hodous: A Dictionary of Chinese Buddhist Terms, p.435a. Rpr. Taipei 1972. TT 1220: 83.6a-6b (*lien-fa* 鍊 法) shows how the internal and spiritual self-identification with and self-deification as Heavenly Mother is achieved in preparation for some Thunder rituals.

18   It literally says: "the repository/intestine spleen". See TT 1250 *Ch´ung-hsü t´ung-miao Shih-ch´en Wang hsien-sheng chia-hua* 5b, for the identification of the element „earth" as a spirit emissary ("The Will") and the "Comprehensively Inspecting Astral Agent of the Great Dipper". See below, TT 1220: 69.8a "the Divinity of the Will" (i-shen 意 神).

Marshal Hsin (辛帥) belongs to [the element] wood and returns to the liver.[19]

The emissaries (*shih-che* 使者) belong to [the element] earth and return to the ancestral palace (*tsu-kung* 祖宮)[20] to realize the union.

[The divinity Chao] Hsüan-t'an ([趙] 玄壇) belongs to the heart,[21] and the Magic Agent Ma (馬靈官) belongs to the heart.

Marshal Wen (溫帥) belongs to [the trigram] *chen* 震. He returns to the liver and passes through (*li* 歷) the gall. The Magic Agent Wang (王靈官) belongs to the heart.[22] Each of [the emissaries] follows the own [super ordinate] marshal and belongs to [one of] the five intestines.

---

19 Concerning these two names see, for example, TT 1220: 82. 28a: *Lei-t'ing san-shuai hsin-lu, shih-hsi, shih-shih* (雷霆三帥心錄世系事實). For an adaptation of astral realities to the human internal spheres, see e.g. TT 1220: 82. 18a-18b. Here we find the comparison of the heart of man with its seven holes and the Northern Dipper (seven stars). Also see Li Yüan-kuo: "*Tao-chiao lei-fa yen-ko k'ao*". In: *Shih-chieh tsung-chiao yen-chiu* 2002, 3, p. 95 (jen-shen hsiao t'ien-ti 人身小天地). Especially see p. 96. See, Liu Chung-yü: *Tao-chiao fa-shu*, p. 84 (i-shen wei i hsiao yü-chou 一身為一小宇宙); and also see F.C.Reiter: "*The Discourse on the Thunders*, by the Taoist Wang Wen-ch'ing (1093-1153), in: JRAS 14/3, pp.210,, 224-225 (2004). Compare TT 1220: 81.1a-16.

20 "ancestral palace" most likely stands for the "central palace".

21 This is Chao Kung-ming (趙公明), a famous name in the Taoist history of the divine, see TT 1476 *Sou-shen chi* 4.10a-11b. The title is Chao Yüan-shuai (趙元帥); abbreviating Cheng-i hsüan-t'an yüan-shuai (正一玄壇元帥). He is said to have avoided the Ch'in (秦) administration and retired to the mountains. See in this book Chapter II, concerning the Altar of Earth Thunder. As to TT 1476, see J.Lévi/F.Verellen, pp.897-989, in: *Companion*. The name of the divinity appears in many Taoist sources. See, for example, TT 1016 *Chen-kao* 10.17b which shows him to be a divinity of the five directions. Consequently see TT 1220: 135.19b (Wu-fang tu-mo shih che Chao Kung-ming 五方都魔使者趙公明). The name Chao Kung-ming meanders through Taoist sources of any epoch. Concerning TT 1016, see I.Robinet, pp.198-200, in: *Companion*. This is an example for the age of the tradition and the vast diffusion of the divine names in our texts. I do not intend to document all these names.

22 Concerning Wang, see F.C.Reiter: "Some Notices on the Magic Agent Wang (Wang *ling-kuan*) at Mt. Ch'i-ch'ü in Tzu-t'ung District, Szechwan Province". In: ZDMG 148, 323-342 (1998). TT 1221 *Shang-ch'ing ling-pao ta-fa* 7.8a-9a offers a cosmic set of 81 Magic Agents, as part of a tradition of 81 amulets. The Agent poses in a prominent position (chu-shuai 主帥). He is second only to the perfected man and spirit patron (chu-fa 主法) Sa Shou-chien (薩守堅), in accordance with the ritual instructions *Lei-t'ing san-wu huo-chü ling-kuan Wang yüan-shuai pi-fa* (雷霆三五火車靈官王元帥秘法), in TT 1220: 241.1a-15b, and see ch. 242 and ch. 243 of TT 1220. In TT 1220: 86.7a the deity has the title Hsien-t'ien san-wu huo-chü ling-kuan Wang Shan (先天三五火車靈官王善). In the same chapter (TT 1220) on p.7b we find the already familiar (see above note 21)

Summoning and Uniting [the Thunder Divinities]   21

When you speak the spells by heart (*nien-chou* 念咒) to summon and unite [with the divinities], they first return to the respective intestine and enter then the central palace. [The orifices] above and below are blocked and they do not allow [divinities] to enter and take lodge. You use the nostrils to exhale noisily, or perhaps you use the sword-*mudrā* (*chien-chüeh* 劍訣) [23] to lead them on to come forth. In front of the altar you can dispatch [the divinities] according to your will and purpose. The rituals of all grades definitely require to unite with and refine [the spiritual forces], or to refine them in meditation and visualizing them when they are present and the inside and the outside unite. You exert your mind and activate the meditation. What follows then are responses that come up like echoes.

The [element] earth is the head of the five elements. Earth lets the essences and the breaths circulate and thoroughly flow through the bones and the nine orifices.

The spleen (*p'i* 脾) is *yin*-earth (陰土). The stomach is *yang*-earth (陽土). The spleen digests the five cereals and connects them to become eminent blossoms (*ying-hua* 英華) that nourish the body. [24]

The stomach (*wei* 胃) is the belly of the human body where all the strings are tightly bundled. It is for this reason **(6a)** that [the stomach] plays the leading part among the five intestines.

As to the character One (一) of the [character] centre (*chung* 中), there is a One *yi* (一) [stroke] at the four sides, and the very centre of [the character] mouth (*k'ou* 口) shows yet another *yi* (一) that is set up in the vertical direction, emerging and reaching upwards. This is called, the divine radiance (*shen-kuang* 神光) rushes towards heaven. Protruding but reaching downwards, [the stroke *yi* 一] is said to be the divine radiance that opposes earth. Now, this means, "the four sides encircle the ultimate". Concerning the character One (一) of the [character] palace (*kung* 宮) that is above or on top of the [character], the One (*yi*

Lei-fu fa hsieh-wu chu-wei hsüan-t'an Chao Lang (雷府伐邪巫朱煒玄壇趙朗) who has a military capacity. Here we also find, for example, Wen Ch'iung (溫瓊/溫帥). This late tradition (*Hsien-t'ien lei-ching yin-shu* 先天雷晶隱書) uses the fundamental concept of *Anterior Heaven* to lay a claim to names that we usually are inclined to combine with the notion of *Posterior Heaven*. Obviously, there are no exclusive categories.
23 See H.A.Giles: A Chinese-English Dictionary. Nr.3225, p.395b. (Rpr. Taipei 1972).
24 Compare TT 432 *Huang-t'ing nei-ching wu-tsang liu-fu pu-hsieh t'u* 13a-14a, also speaking about the stomach. For this text (the preface is dated 848), see J.Lévi, in: *Companion*, pp. 348-349.

一) turns out to be the cover. The upper [character] mouth (口) is the tripod vessel (鼎 器) that keeps its orifice empty as to store inside the substance of the subtle shining of pure gold that is the ancestral breath (tsu-ch'i 祖 氣).

The one lower [character] mouth (口) [of the character kung (宮)] is the square stove that puts in order and accomplishes the tripod vessel. Its upper part is round and resembles heaven. Its lower part is square and resembles earth. Heaven is ch'ien (乾), and the earth is k'un (坤). This is called the orifice of the heavenly female (hsüan-p'in 玄 牝). [The word] heavenly (hsüan 玄) has the meaning of black and heaven. [The word] female (牝) has the meaning of yellow and earth.[25] Earth is the caldron (fu 釜), and the caldron is the valley. The valley is an empty vessel that holds inside the original breath (yüan-ch'i 元 氣)[26] and collects the essences and the spiritual forces. They are aroused by thunders and thunderclaps. They are fertilized by wind and rain.[27] Just everything is contained herein.

[Wang Wen-ch'ing (王 文 卿) continues saying:] The elixir classics (tan-ching 丹 經) say: "When the one yang (一 陽) agitated for the first time, where did the agitation occur? When the black crow soared up, from which spot did it arise?[28] **(6b)** Now, it all came from the point of agitation of the one spark, which was the time of the absolute beginning of the greatest ultimate (t'ai-chi 太 極). When that one spark had not yet agitated, my own body and the breaths of heaven and earth mutually were the inside and the outside.

The commentary by Yü-feng (御 風) says: The one (一) is yang (陽); and the one is pure ch'ien (ch'un-ch'ien 純 乾). Ch'ien (乾) is the perfect yang (chen-yang 真 陽). Yang gives birth to offsprings (tzu 子). At the very beginning, the one yang is

---

25 Compare the explanation of the term hsüan-p'in (玄 牝) in the Ho-shang kung (河 上 公) commentary TT 682 Tao-te chen-ching chu 1.5a (ch'eng-hsiang 6/ku-shen pu ssu 成 象 / 谷 神 不 死), which may suggest that Wang Wen-ch'ing (王 文 卿) carried on a certain line of tradition. A good example for the grossly diverging interpretation of the heavenly or mysterious female as the absolute essential basis that is beyond any wording, gives, for example, TT 263 Hsiu-chen shih-shu tsa-chu chih-hsüan p'ien 5.4a-4b; and ch.9.3a (Chin-tan ta-ch'eng chi/ Hsüan-p'in t'u 金 丹 大 成 集 玄 牝 圖).
26 See TT 1250: 14a, the original breath is a bodily reality.
27 Chou-i 周 易 17, hsi-tz'u hsia 繫 辭 下 1, p.2. See, The I Ching or Book of Changes, the Richard Wilhelm Translation rendered into English, by C.F.Baynes. New York 1967. (Henceforth: Book of Changes). See, pp. 283, 284 where the word "lightning" is used instead of "thunderclap". Wang Wen-ch'ing (王 文 卿) reverses the sequence of the phrases in the I-ching (易 經).
28 The black crow represents the sun.

most enduring and sound. There is just nothing that is mixed up with it. [The one *yang*] is entrusted to the womb at the centre of *k'an* (坎).²⁹ When the human being comes to life for the first time, the one *yang* (陽) had first been agitated. When father and mother wed their essences, then it is that heaven and the one (*t'ien-i* 天 一) ³⁰ give birth to water, and so it is that [the element] fire is born within water. The fire is *yang*. When *yang* agitates, then it is that the very being streams [forth] (*hsing-liu* 性 流) to the female creatures (*p'in-wu* 牝 物) that manifest their being. They definitely come to life due to heaven and the one. Water is goodness (*jen* 仁). Goodness means, "to be ripe" (*shih* 碩). The inside of a fruit produces a kernel, and what is stored inside the kernel, this is the goodness. Goodness, this is the perfect water (*chen-shui* 真 水), and water was brought to life by heaven and the one (*t'ien-i* 天 一). When [this] agitation (*tung* 動) occurs, the *yang*-breath (*yang-ch'i* 陽 氣) ascends and rising up forms high above the blissful clouds (*ch'ing-yün* 慶 雲) that open the door of life (*sheng-men* 生 門) and let auspicious vapours flow down to block the window of death (*ssu-hu* 死 戶). Such utmost subtle, such utmost mysterious **(7a)** principles they are! High above there is the bent *chiang*-river (江). See the moonflower that is flourishing and pure, and there is the black crow that soars high above. This is something that genuinely belongs to everybody. When amidst great tranquillity a [specific] sensation (*hsiao-hsi* 消 息) arises, the one spark (*i-tian* 一 點) [of agitation] comes forth, issuing from where the Anterior Heaven (*hsien-t'ien* 先 天) has not yet started to bud. You must know the utmost beginning when father and mother had not yet come to life. There was the situation that the one *yang* (陽) first agitated. At that time when all creatures had not yet come to life, they were without the resources to grasp life. It is no wonder that the bodies received for the first time their vital breaths due to the initiating power of heaven (*t'ien-chi* 天 機) ³¹ that completely leaked out. These words say it all.

---

29 This refers to the trigram *k'an* that has one *yang*-stroke in its very centre; see *Book of Changes*, p.114 (hexagram *k'an*, *The Abysmal/Water*).
30 TT 1250: 11a, for the combination of *t'ien-i* (天 一) with the element "water" and the gall-[department], see below. TT 1250: 11a-11b, gives the context of an internal practice, which is part of the process of creation that is spiritually realised within the human body.
31 TT 1250 *Ch'ung-hsü t'ung-miao Shih-ch'en Wang hsien-sheng chia-hua* 7b-8a, for a definition and description of the „heavenly initiating (or moving) power" (*t'ien-chi* 天 機) in terms of cyclic astronomic processes during the year, involving the 28 stellar mansions, the 24 phases of the breaths and the stimulation of the 36 shining elements. TT 1250: 6a shows how the "moving power" can be manipulated following to the application of appropriate amulets.

[Wang Wen-ch'ing (王文卿) continues saying:] The spleen (p'i 脾) is [called] Furnace Crescent Moon (yen-yüeh lu 偃月爐).[32] Its outer shape shows two horns (chiao 角) that are sharp and hang down. At the one spark of the first agitation, the two horns bend upwards. Between the two horns, there is one orifice, which has the size of a black corn of millet. When the one orifice is about to open, the time has come for the internal practice [of self-cultivation] (hsing-ch'ih 行持). You visualize befittingly the one spark of golden radiance that comes forth from the cave [that resembles] a black corn of millet.

In case that you summon the general of the kidneys, it is necessary that this one spark of gold radiance be properly attached to the kidneys. In case that you summon the generals of the heart and the gall, you should concentrate (ts'un 存) on this one spark and attach it to the heart and the gall. The accompanying [spiritual] forces can then be deployed (yun-hua 運化) when that [procedure is complete]. **(7b)** [The ritual of] summoning and uniting works just like that, and there will never be any failure.

The commentary by Yü-feng (御風) says: The spleen is the Furnace Crescent Moon that [resembles] an iron sickle. The head is big and the tail is sharp, being long five ts'un (寸). It is also called the Perfect Earth of the Five Directions. The spleen is tightly connected with the heart. The orifices of the heart irrigate the eyes and the eyebrows, and they irrigate [the element] earth of the spleen. All the ritual officers[33] who summon and dispatch the [Thunder] generals keep their two eyes looking upwards, and they force the two eyebrows to bend upwards. A sustaining force is thus raised high above across the heaven. Then it is that the heavenly orifices (hsüan-ch'iao 玄竅) issue spontaneously the apparition of a divine radiance, which supports heaven. The thunder divinities (lei-shen 雷神) can hardly afford not to respond. The ancestral breaths (tsu-ch'i

---

32 TT 1220: 69.25a, connecting the spirit force of the spleen (p'i-shen 脾神) with the will (i 意), the element earth (t'u 土) and the numbers five and ten. However, the image of the furnace does not appear again. TT 1402 Shang-ch'ing huang-t'ing wu-tsang liu-fu chen-jen yü-chou ching 6b-8a compares the spleen with an "overturned bowl". For this T'ang-Text see J.Lévi, in *Companion* pp.350-351. For a Furnace Crescent Moon" (yen-yüeh chih lu 偃月之鑪) serving as a metaphor for the area of internal refinement being compared with the Tripod of Red Cinnabar (chu-sha chih ting 朱砂之鼎), see TT 263 *Hsiu-chen shih-shu tsa-chu chih-hsüan p'ien* 6.2b; also see ch.9,2b (t'ien-hsin t'u 天心圖). See TT 263 *Hsiu-chen shih-shu wu-chen p'ien* 26.7a, which shows a drawing of a Furnace Crescent Moon, which may suffice here as documentation.

33 *Fa-kuan* (法官) should refer to the acting priest (tao-shih). TT 1250: 13b has the combined term "ritual officers and Taoist priests" (fa-kuan tao-shih 法官道士) which in this case, however, should not point to persons of different rank and professional quality.

祖 氣), however, are the golden radiance of the Anterior Heaven (*hsien-t'ien* 先 天) that shines brightly and since antiquity was never extinguished. The *Tu-jen ching* says: "It was in the middle of the void heaven that [Yüan-shih (元 始)] suspended one precious pearl as small as one grain of millet. It was separated five *chang* (丈) from the earth." [34] Earth belongs to the trigram *li* (*li-kua* 離 卦). This trigram has three strokes. Two strokes, the upper one and the lower one are *ch'ien* (乾) and metal (*chin* 金). In the middle between the two strokes, there is *k'un* (坤). *K'un* is the earth (*ti* 地), and the earth **(8a)** is *yin*-earth (*yin-t'u* 陰 土). [The element] earth (土) constitutes the earth (*ti* 地). Moving away from it five *ts'un* you touch the very centre (*chung-huang* 中 黃). The upper orifices of the eight poles are *ch'ien* (乾) and the yellow of heaven (*t'ien-huang* 天 黃). The yellow of heaven is exactly [identical with] the mysterious yellow (*hsüan-huang* 玄 黃). The lower orifices are the earth and the yellow of the earth (*ti-huang* 地 黃). Yellow, that is the yellow court (*huang-t'ing* 黃 庭). The yellow court is a room (*wu* 屋) and its base does not move (*tung* 動). The three orifices of its right side [correspond with the trigrams] *chen, k'an* and *ken* (震 坎 艮), which are the three males (*nan* 男). The three orifices of its left side [correspond with the trigrams] *sun, li* and *tui* (巽 離 兌), which are the three females (*nü* 女). *Ch'ien* (乾) and *k'un* (坤) together give birth to six children. The six children circulate and move on day and night without any pause. The movements of the heavenly bodies are strong by themselves and go on without any pause. Humans and quadrupeds all possess [internally] this most subtle principle. As soon as the eyes, the ears, the mouth, the hands and feet rise and act, these orifices also move. The [related] subtle mysteries are subject to oral transmissions (*k'ou ch'uan* 口 傳). Only the quadrupeds carry their heart in a horizontal position, and therefore they are born as domestic animals. Birds, however, do not have these orifices. There are

---

34 See TT 147 *Ling-pao wu-liang tu-jen shang-p'in miao-ching fu-t'u* 1.6b. See TT 1250: 4b, which correlates the Anterior Heaven with the Original Spiritual Force and the Prime Origin that "nowadays are venerated" (yüan-shen 元 神 / yüan-shih 元 始). For a Thunder ritual that completely is dedicated to the Anterior Heaven and its unified breaths, see TT 1220: 90 (*Hsien-t'ien i-ch'i lei-fa* 先 天 一 氣 雷 法) that incorporates the famous Song of the Heavenly Pearl by Wang Wen-ch'ing: *Shih-ch'en hsüan-chu ke* 侍 宸 玄 珠 歌; 90.15b-16b). Compare L.Skar: "Administering Thunder: A thirteenth Century Memorial Deliberating the Thunder Rites", in: *Cahiers d'Extrême-Asie* 9, pp.159-202 (1996-1997). The Anterior Heaven matches the Anterior Earth (hsien-ti 先 地) in *Hsüan-chu ke* (玄 珠 歌) by Wang Wen-ch'ing (王 文 卿), see TT 1220: 70.4b. Also see, for example, TT 1221 *Shang-ch'ing ling-pao ta-fa* 40.19b (*Pi-lo k'ung-ko fu* 碧 落 空 歌 符), for the self-identification or self-divinisation of the acting priest with Yüan-shih [t'ien-tsun] (元 始 天 尊). For the song *Pi-lo k'ung ko* (碧 落 空 歌) and the appropriate chart, see TT 147 *Ling-pao wu-liang tu-jen shang-p'in miao-ching fu-t'u* 1.3a-3b (the preface was done by Sung Hui-tsung 宋 徽 宗). Compare J.Lagerwey, in: *Companion* pp.1084-1085.

some most subtle oral instructions, and the mysteries refer to the heart. Keep them for you as a secret, keep them secret!

In case that you summon the [spirit] emissaries, you take the divinity of the will (*i-shen* 意 神) to be the leading force that is *lien-chen* (㢘 貞) [35] which is the internal *kang* (罡) –star. [The star] becomes apparent due to its **(8b)** golden radiance.

In case that you summon Marshal Teng (Teng shuai 鄧 帥), you take the heart to be the fire official (*huo-kuan* 火 官). The fire official is the heart, and the heart is the heavenly *kang*-star (*t'ien-kang* 天 罡). [36] When the heavenly *kang*-star shakes, golden radiance concentrates and shoots forth from the [inner] central palace (*chung-kung* 中 宮). How could it happen that Marshal Teng would not become magically efficient (*pu-ling* 不 靈)?

In case that you summon Marshal Hsin (Hsin shuai 辛 帥), you take the divinity of the breath (*ch'i-shen* 氣 神) to be the leading force. It belongs to the liver, passes through the gall department (*tan-fu* 膽 府) and flows into the central palace (中 宮) where golden radiance gushes forth and scatters. [37]

In case that you summon Marshal Wen (Wen shuai 溫 帥), you make use of Wood Senior (*mu-lao* 木 老). When [the element] fire is born for the first time it is based on the rage of the liver. It happens then that Marshal Wen makes his apparition.

Wood gives birth to the fire of the heart (*hsin-huo* 心 火), and fire is the divinity *ping-ting* (丙 丁) [38] that makes the heavenly *kang*-star (*t'ien-kang* 天 罡) shake and move. The golden radiance in the central palace resembles the shape of flowing brass that is getting mixed and refined. The [related] mysteries will be exposed in

---

35 See F.C.Reiter: "The Discourse on the Thunders 雷 說, by the Taoist Wang Wen-ch'ing 王 文 卿 (1093-1153)", in: *JRAS* 14/3, p. 220.
36 Compare TT 1220: 70.1b (*Hsüan-chu ko* 玄 珠 歌, commentary by Bai Yü-ch'an 白 玉 蟾: "*t'ien-kang hsin yeh* 天 罡 心 也"). TT 1220: 82.18a-18b features the *t'ien-kang* [star] (天 罡) as an object of individual internal possession: "my heart is inside of me but outside of me it is the k'ang-chen [star] (㢘 貞). See TT 1220: 151.3a-5a (*t'ien-kang fa/t'ien-kang shuo* 天 罡 法 / 天 罡 說), for astronomical interpretations.
37 Both Marshals (Teng shuai 鄧 帥 and Hsin shuai 辛 帥) patronize their own sets of rituals, see TT 1220: 80.1a-45b (*Yen-huo lü-ling Teng t'ien-chün ta-fa* 焱 火 律 令 鄧 天 君 大 法); TT 1220: 81.1a-16b (*Fu-feng meng-li Hsin t'ien-chün ta-fa* 負 風 猛 吏 辛 天 君 大 法).
38 This addresses most likely the heat [of the sun] (*ping-ting [huo]* 丙 丁 [火]), see below the translation of TT 1220: 56.22b.

oral instructions, and each of them is to be applied in accordance with the [respective spirit] general or marshal.

[Wang Wen-ch'ing (王文卿) continues saying:] The *Hung-fan* (洪範) says: "Perspicacity manifests itself in wisdom (*sheng* 聖)".[39] These words are exactly to the point and pertain to all ranks. In a ritual, you must desire to perform the internal self-cultivation (*hsing-ch'ih* 行持).[40] You should befittingly stop up your own complete breath (*wu i ch'i* 吾一氣), and then it is that all worldly causes (*yüan* 緣) cease to be and any thoughts do not arise. **(9a)** This exactly is your great goal.

The commentary by Yü-feng (御風) explains: The *Book of Documents* says that perspicacity manifests itself in wisdom. These words are absolutely to the point. Accordingly, the breath of greatest purity within my body is kept in store at the heavenly joints (*hsüan-kuan* 玄關).[41] The central breath of the Great Ultimate is the principle (*li* 理), and the principle gives birth to being (*hsing* 性). Being gives birth to the dragon of *yang* (陽龍), and the dragon of *yang* is able to transform (*pien-hua* 變化) and ascend (*sheng-t'eng* 昇騰). The cosmos is within your hands exactly at this moment, and all mutations come to life [within your own] body.[42]

[Wang Wen-ch'ing (王文卿) continues saying:] Now, coming to the actual site of [a ritual of] Summoning and Uniting, the [following] question may be put forth: At the very time of Summoning and Uniting, how can I find out whether the divine generals have arrived at the scene or not? The answer is that you must definitely know the coming or the going [of the divinities]. If you do not know whether the divine generals have arrived or not, this is called to perform a blind ritual (*hsia-fa* 瞎法).

The commentary by Yü-feng (御風) says: When the spiritual force agitates, the breath follows suit. [The term] spiritual force (神) means the original spiritual

---

39 Huang K'an ed.: *Pai-wen shih-san ching, Shang-shu; Chou-shu hung-fan* p. 34/4. Shanghai, Ku-chi Comp. 1983.
40 This term points to chapter three in this text that uses the phrase as its title, see TT 1220: 69.11a-14a.
41 This may well refer to the cinnabar fields. However, we do not get any conclusive annotation on this term by Wang Wen-ch'ing (王文卿) or his commentator. Concerning a combination of external cosmic and internal bodily dimensions of this term see, for example, the spell *Shih-erh ching-lo fu-chou* (十二經絡符咒), in TT 1221 *Shang-ch'ing ling-pao ta-fa* 51.13a.
42 Compare TT 31 *Huang-ti yin-fu ching* 1a (*Shen-hsien pao -i, yen-tao* 神仙抱一演道). The dragon of *yang* appears to be a *pars pro toto* representing the respective practitioner.

force (元神).[43] The original spiritual force is precisely the individually owned original *yang* (*yuan-yang* 元陽). The original *yang* is the breath and the same time the principle. The principle gives birth to being (性). Being gives birth to the dragon of *yang*. **(9b)** The dragon of *yang* is the torch of wisdom (*hui-chu* 慧燭). At the time of Summoning and Uniting the nostrils inhale the breath of pure *yang* (*ch'ing-yang* 清陽) of the Anterior Heaven (*hsien-t'ien* 先天), and you transfer that breath to return to the yellow court (*huang-t'ing* 黃庭) where [the breath] congeals. Being tranquil for a short moment [the breath] must then agitate, and the divine generals have arrived when [the breath] agitates.

[Wang Wen-ch'ing (王文卿) continues saying:] As to the coming [of the divine generals], at the time when they are summoned and united you must be silent and tranquil.[44] Perhaps you sense [their presence] when both of your temples feel a cold as if a cold wind had arrived or as if [cold wind] had touched down on both of your temples. Perhaps it is that your eyes flicker or your nostrils ache or perhaps there is the sound of bells in your ears, or perhaps it is that wind and thunders drum and agitate. In all these cases, [we know that] the heavenly generals (*t'ien-chiang* 天將) have arrived.

The commentary by Yü-feng (御風) says: As to the coming [of the spirit generals] at the time of Summoning and Uniting, when in the state of tranquillity a sensation occurs you first speak the spell *San-ching chou* (三淨祝). You use the nostrils to inhale noisily breath to fill your mouth. Using the nostrils, you lead the breath gently into the central palace (*chung-kung* 中宮). Wait until there is a true sensation (*chen hsiao-hsi* 真消息) between the two kidneys. [You remain] in greatest tranquillity like being frozen and motionless, and so you quietly listen where the original spiritual force agitates on its own and where it stays at. Then it is that the heavenly generals **(10a)** have actually arrived. Never mind which general it is, they all start out to operate at the central palace.

[Wang Wen-ch'ing (王文卿) continues saying:] At this moment, you make one loud cry, pull aside, and turn the handle of the dipper (*hsien-fan tou-ping* 掀翻斗柄). You most urgently press the *mudrâ* Thunder Office (*lei-chü* 雷局)[45] of your left hand into the hip. The left foot treads heavily on the ground. You concentrate [your vision] on the general that was summoned and right now stays

---

43 TT 1250: 3a identifies the original spiritual force with the very body and the very being of the thunders (*lei chih t'i-hsing yüan-shen yeh* 雷之體性元神也).
44 Compare TT 1250: 10b.
45 See TT 1250: 3b, for an application of the Thunder Office. The word *chü* (局) can mean a position that holds a potential. I always use Thunder Office in this sense as a name.

above the window of earth (*ti-hu* 地 戶).⁴⁶ Perhaps you draw an amulet and send (*t'i* 提) the general to enter the amulet, or you dispatch [the general] to execute your ritual orders (*hsing-shih* 行 事). However, when you urgently speak a spell to dispatch [a spirit general] concerning any ritual task, you must not relax (*fang* 放) [the *mudrâ* of] the left [Thunder] Office and you must not move away your left foot because [otherwise] the [divine] general is gone.⁴⁷ In case that you summon once but the [spirit] general does not respond, then it had happened that the heavenly general did not come. You must repeat the summoning. In case that after three summonses there is still not any response at all, the respective general just does not descend. In this case, however, any ritual performances will not have any magic results. My intentions incline to focus on this secret matter, which I must not put into words. Be cautious! Be cautious!

The commentary by Yü-feng (御 風) says: Shouting out aloud at one time, you alert the generals and [their] emissaries to pull aside and turn the handle of the dipper, **(10b)** and then the Northern Dipper (*pei-tou* 北 斗) faces southwards. Notice, the heart has seven orifices that are the Northern Dipper. The hair [of the eyebrows] is the Three Terraces [-stars] (*san-t'ai* 三 台).⁴⁸ The nostrils inhale the breath of pure *yang* of the Anterior Heaven (*hsien-t'ien* 先 天) and receive [the breath] to return [to your body]. The orifices of the heart are tightly shut down and do not allow the breath to be contained there but let it continue to flow up to the whole face with its seven apertures (*ch'i-k'ung* 七 孔) until it flows

---

46 This is the direction of *sun* 巽, which points to the Southeast. Compare TT 1250: 4a.
47 ...and no longer is available for ritual service.
48 San-t'ai (三 台), this is the "flower baldachin", the three stars above the dipper. See TT 1220: 58.1b, for a drawing of the Big Dipper, explicitly showing the stars San-t'ai (三 台). Compare TT 1227 *T'ai-shang chu-kuo chiu-min tsung-chen pi-yao* 2.12b (san-t'ai hsing-hsing 三 台 星 形). For another interpretation see P. Andersen: "The Practice of *Bugang*", in: *Cahiers d'Extrême-Asie* 5, p.30 (1989-1990). The term *mao* (毛) is not very clear but may refer to the eyebrows. For a central function of the stars San-t'ai (三 台) in a spell that is connected with a *Huo-sha* amulet (火 煞 符) see TT 1220: 139.4b; also see, for example, TT 1221 *Shang-ch'ing ling-pao ta-fa* 5.15b, for a very instructive spell that focuses on the creative capabilities of the stars San-t'ai (三 台). TT 148 *Wu-liang tu-jen shang-p'in miao-ching p'ang t'ung-t'u* 2.1a refers to the astronomical monograph *Shih-chi t'ien-kuan shu* (史 記 天 官 書) as to explain the superb function of the Big Dipper ("carriage of the God-Emperors" ti-chü 帝 車) in organising the circuit of the time and the seasons. Also see Liu Chung-yü: *Tao-chiao fa-shu*, p.183, with an alternative name (*san-neng* 三 能) and a greatly differing description and identification of the term. Such terms are crystallisations of a vast array of regional and historical traditions that are beyond any certain identification. Today, the three dots that top many amulets are explained in present day Taiwan to be (from the left) the lower, middle and upper platform stars, suggesting the formation of a triangle.

down to the belly and the colon (*ku-tao* 谷 道). The breath thus urges on for a long time.

The nostrils inhale the breaths that tightly bind up the two kidneys and ascend from the spine (*chia-chi* 夾 脊). This exactly is [the meaning of] "pulling aside and turning the handle of the dipper" (*hsien-fan tou-ping* 掀 翻 斗 柄). Urgently take [the *mudrâ*] Thunder Office of the left hand and the sword [-*mudrâ*] (*chien-chüeh* 劍 訣) of the right hand and press them into your hips.[49] Your left foot treads heavily on the ground, and you meditate on the general and the emissaries that were summoned as they arrive at the window of earth. Perhaps you draw an amulet and dispatch a general. You use your mouth to inhale [him] and blow [him] out to enter the amulet, or you dispatch the [spirit] general to take care of ritual matters. You speak then the spell that suits the occasion. When you dispatch [generals and emissaries] in this way, you must definitely not release the *mudrâ* Thunder Office of the left hand, and you must not move away the left foot, because otherwise the [divine] general would be gone. In case that you summon one time [but the divinity] does not come, and when the same happens a second time and a third time, the [divine] general just does not descend, and any application of amulets [50] in order to cure illness (*chih-ping* 治 病) **(11a)** definitely does not have any magic result. ***

The text combines a welter of information about the connection and union of the priest with the divine entities that he may summon. Many thunder divinities go unnamed in our texts.[51] However, in the text Summoning and Uniting [the Thunder Divinities] (*chao-ho* 召 合) we see a selection of names of divine figures that often appear in texts that are attributed to Wang Wen-ch´ing (王 文 卿). We certainly notice that much of the extensive and detailed expositions in terms of internal alchemy (*nei-tan* 內 丹) were phrased by Yü-feng (御 風) who is an otherwise unknown disciple of Wang Wen-ch´ing (王 文 卿). The expositions of internal alchemy are as much specific as they are quite general. They represent a certain mode of theoretical approach. Sung-specialists of internal alchemy used

---

49 TT 1220: 69.25a (commentary by Yü-feng 御 風 on Wang Wen-ch´ing´s (王 文 卿) tract The Creative Impetus (*Tsao-hua* 造 化) and see, for example, TT 1220: 85. 8b: "tread on and turn the handle of the dipper" (t´a-fan tou-ping 踏 翻 斗 柄). Compare a later tradition in TT 263 *Hsiu-chen shih-shu* 3.5a (*Yin-fu sui* 陰 符 髓). Concerning the *mudrâ chien-chüeh* (劍 訣), see also TT 1220: 68.1b, in combination with the *mudrâ* Thunder Office.
50 Concerning this important theme, see Chapter II, and also see, for example, Li Yüan-kuo: "*Lung tao-fu te chieh-kou yü pi-fa*" (論 道 符 的 結 構 與 筆 法), in: *Tsung-chiao hsüeh yen-chiu* 1992, 2, pp.8-13.
51 See F.C.Reiter: "The Name of the Nameless and Thunder Magic", pp.97-116, in: AAS 20.

such and similar descriptions of energetic dispositions within the human body that seemed to mirror cosmic and astral spheres. Obviously, there is some deliberately cryptic imagination involved. On the other hand, we see clearly how internal alchemy is being applied in rituals of Taoist Thunder Magic.

Wang Wen-ch´ing (王文卿) presents rather scarce and basic information, whereas his disciple Yü-feng (御風) rephrases and enlarges the frame of interpretation due to his understanding. It is rather seldom that this type of rhetoric appears in other sources of Thunder Magic. Obviously, different levels of rhetoric and presentation were used to explicate the fabric of Thunder Magic. We have many indications that the internal self-cultivation was understood to be crucial. [52] We also find out that such explicit and extended information on Taoist self cultivation is usually contained in later texts that emerged in the school of Pai Yü-ch´an (白玉蟾 13th ct.) when the development of internal alchemy (*nei-tan* 內丹) peaked.

The very first text in this presentation, Secret Instructions Concerning the Rituals of the Thunders (*lei-fa pi-chih* 雷法秘旨), featured the self-identification and self-deification of the Taoist who in this way prepares himself to perform a Thunder ritual. The ritual task is the sole incentive for any self-cultivation. It is all about the purpose of preparing the human body and person to be fit and ready for the ritual performance. In fact, the self-deification is already a ritual in itself. We shall see that this is a great theme in Taoist canonical texts, which specialise on exorcist rituals and almost completely forgo any rhetoric or didactic explanations and justification. [53] However, to support the line of reasoning that we have seen so far I introduce another text by Wang Wen-ch´ing (王文卿) that shows how to employ self-cultivation for ritual purposes. This text substantiates the information that the Secret Instructions Concerning the Rituals of the Thunders (*lei-fa pi-chih* 雷法秘旨) already have conveyed.

---

52 See, for example, TT 1220: 76.39a-39b (Preface by Pai Yü-ch´an 白玉蟾).
53 See Chapter II: The Scope of Thunder Magic

Assembling the Divine Force
(Lien-shen 鍊 神) [54]

(2a) Let your divine forces congeal and sit quietly in meditation. You concentrate on the one most shining point in the Kidney Palace (*shen-kung* 腎 宮). Within a short moment, fire arises, gradually engulfing your body all around. You blow out one load of breath from your mouth, and the ashes will be blown away altogether. Then, you concentrate on the breaths in the five colours of the five directions, which mix and combine to shape one united aura of radiant shining in purple and golden colours, and this [aura] transforms itself into an infant (*ying-erh* 嬰 兒) that gradually grows big. [This image] has the beak of a phoenix with silver teeth, red hair and a body like a quail. Both eyes let fiery shining penetrate [a distance] of ten thousand *chang*. Both wings also have [the shining of] fire. On both forelegs, a head with eyes emerges. Each of them also emits fiery shining. The belt has the colour of gold. The left hand clutches a fire auger and the right hand clutches a mallet with eight angles. A fiery dragon winds around the body.[55]

Thereupon you concentrate and see yourself as this Divine General of the Five Thunders. His head touches the heaven, and he stands on the earth. Close around him there are fiery clouds that wrap him up with the divine and fierce might of blazing fire. This is "Blazing Fire", the Heavenly Lord Teng (Yen-huo

---

54 TT 1220: 124.1b-2a, for a translation with very few variants see F.C.Reiter: "A Preliminary Study of the Taoist Wang Wen-ch'ing (1093-1153) and his Thunder Magic (*lei-fa*)", in: ZDMG 152, p.172 (2002). This tract shows that Wang Wen-ch'ing (王 文 卿) is embedded in the Taoist traditions of his time, quite apart from the specifications of Thunder Magic and the respective deities that Thunder Magic addresses. For example, compare TT 1221 *Shang-ch'ing ling-pao ta-fa* 54.20b-21b (*lien-hsing hsing-ch'ih* 煉 形 行 持). TT 1220: 124.1b-2a contains the text as a preliminary instruction for the Thunder rituals *Shang-ch'ing lei-t'ing huo-chü wu-lei ta-fa* (上 清 雷 霆 火 車 五 雷 大 法), for which we also get a preface by Wang Wen-ch'ing (王 文 卿). This proves that we have to deal with independent sets of rituals that partly were grouped around the name of Wang Wen-ch'ing (王 文 卿). For the same story see TT 1220: 56.14b-15b, and below Chapter II.
55 The text TT 1220: 80.1a-1b *Yen-huo lü-ling Teng t'ien-chün ta-fa* (焱 火 律 令 鄧 天 君 大 法) shows a very good example for later (probably 13th ct.) *addenda* and embellishments of the status symbols of this divinity. The divinity has, among some other characteristics, "three eyes", and below the two wings, there are "two heads. The left one is in charge of the wind, and the right one is in charge of the rain. The whole body of the divinity is engulfed in fierce fire, riding a red dragon". There is no exclusive canon of such marks of identity, which religious imagination freely moulds and enlarges on the basic pattern of the body of a quail.

Teng t'ien-chün 焱 火 鄧 天 君) who is the ruling and commanding divinity in the rituals of the fire chariots (Huo-chü fa 火 車 法)."

The priest transforms himself to be the Heavenly Lord Teng [56] to adopt the divine capacity of the divinity. The ensuing ritual action is not perfomed on behalf of that deity but, in other words, the deity is the performing agent. This is the crucial point, which generally is the very basis for any Thunder Magic rituals. In this case, the meditating priest creats out of his potentials the Heavenly Lord Teng. Here we have to make the point that the deity certainly does not come down from anywhere to possess the priest but is an innate spiritual potential.

The two preceding texts contain some very common and practical elements, for example, the hand-gesture (mudrâ) Thunder Office (lei-chü 雷 局) and the meditative and internal visions that invoke Thunder divinities and the same time spiritually unite the human body with cosmic realities. We also learn about a few names of divinities that are important for the Thunder Magic of Wang Wen-ch'ing (王 文 卿).

The ritual traditions that we have seen so far surface again, for example, in the short commentary on a spell (chou 咒) [57] that has to be spoken within the sacred area, right after two other spells with the title Divine Spell for the Ritual Steps to Turn Divine (Pu-kang pien-shen chou 步 罡 變 神 咒) already had been spoken. The spell invites quite a number of divinities to descend namely the two Marshals Hsin and Chang (辛 張 二 帥). There are the Thunder Lord and the Mother of Lightning (Lei-kung tien-mu 雷 公 電 母), [58] the Wind Earl (feng-po 風 伯) and the Master of Rain (yü-shih 雨 師), the Savage Thunders of the Five Directions (wu-fang man-lei 五 方 蠻 雷) and others.[59] Now, the commentary

---

56 Concerning this name, see also F.C.Reiter: „The Discourse on the Thunders", p.224; and also TT 1220: 80.1a sq.
57 See TT 1220: 87.1b/column 4 sq.
58 Concerning Lei-kung (Lei-shen 雷 神 / 雷 公), see R.Mathieu: Étude sur la mythologie et l'ethnologie de la Chine ancienne, traduction annotée du Shanhai jing vol 1, p.503, and also note 2 (Paris 1983). Especially see T'ai-p'ing kuang-chi (太 平 廣 記) 394 (lei 雷 1-3), pp.1601-1612. See TT 263 Hsiu-chen shih-shu shang-ch'ing chi 39a- 39b (Ch'i-yü ko 祈 雨 歌) makes mention of Lei-kung and Tien-mu (雷 公 電 母). See F.C.Reiter: "The Name of the Nameless And Thunder Magic", p.101, in: AAS 20. Also see TT 1015 Chin-so liu-chu yin 4.10b-11a, naming Lei-kung (雷 公) and referring to the story of Ch'ih-yu (蚩 尤) that Wang Wen-ch'ing (王 文 卿) used as a frame for his well known presentation of the superior Thunder deity Yen-huo ta-shen (焱 火 大 神).
59 Concerning these names, see F.C.Reiter: "The Discourse on the Thunders", pp.210, 224. The names appear again in the commentary that I translate, see below.

that I present shows the combination of ritual gestures and internal processes. Again, we understand that such processes and concepts were right at the basis of Thunder Magic rituals. [60] I present the commentary that refers to the second cryptic phrase of the spell that reads as follows: [61]

(3a) The red [colour] at the red season resembles ferocious blood (*hung-shih hung ssu meng-meng hsüeh* 紅 時 紅 似 猛 猛 血).

The anonymous commentary says: "Your two hands form the *mudrā* Thunder Office (*lei-chü* 雷 局). You concentrate on your tongue as the Thunder Axe (*lei-fu* 雷 斧). The gall is the rumbling [of Thunder] (*p'i-li* 霹 靂). The heart is Marshal Teng (Teng shuai 鄧 帥) **(3b)**. The gall is Marshal Hsin (Hsin shuai 辛 帥). The kidneys are Marshal Chang (Chang shuai 張 帥). The five intestines are the Five Thunders. You concentrate then on Marshal Teng (Teng shuai 鄧 帥) who mounts the red breaths and descends to the Palace of the Kidneys. This is called: the *yang*-breaths (陽) descend. Marshal Chang mounts the black breaths and ascends to the Palace of the Heart. This is called: the *yin*-breaths (陰) storm upwards. In case that such an ascent and a descent happen three times, *yin* and *yang* (陰 陽) strike clashing and become the thunder that enters the spleen. You shut up the breaths and keep the sight of your eyes fixed on the top of your head. You spontaneously lead the breaths of the kidneys upwards into the heart. [Again,] you shut up the breaths and [close] your eyes, and you feel spontaneously how the breaths of the heart descend to the kidneys. At the next step, you concentrate on the Savage Thunders (*man-lei* 蠻 雷) of the Five Directions. Each of them mounts the breaths of the five intestines and they flow altogether into the spleen. The Rumbling Great Divinity (*p'i-li ta-shen* 霹 靂 大 神) together with the [Thunder] emissaries mounts the breaths of the gall. Furthermore, the breaths enter the centre of the spleen where they congeal and combine [as if they were in] a tub. [62] [The breaths] revolve without cease and reach revolving the lower cinnabar field (*hsia tan-t'ien* 下 丹 田). This is called "to mix and unite with the Three Palaces". [63] Repeatedly revolving [the breaths] reach the central cinnabar field (*chung tan-t'ien* 中 丹 田), which is called "the centre organises the five breaths". The five breaths revolve and move on,

---

60 For the commentary, see TT 1220: 87.3a-4a. Also see F.C.Reiter: "The Discourse on the Thunders", pp.222, 224, 228.
61 TT 1220: 87.3a
62 The reading „tub" is a guess on my part, compare H.A.Giles: A Chinese-English Dictionary, nr. 12289 (t'ung 桶). I could not clearly identify the character.
63 The three palaces most likely point to the cinnabar fields, including the Mud Pill (Palace).

forcing their way upwards to the Mud Pill [Palace] (*ni-wan*泥 丸). This is called (4a) "to mix and unite with all the spiritual forces".

It is necessary to use the *mudrâ* Thunder Office (*lei-chü* 雷 局) and to move [the two hands with the *mudrâs*] upwards firmly pressing [them]. You start out doing so from the hall of the two kidneys (*shen-t'ang* 腎 堂), and following the two sides [of your body] you reach the two ears. Now, you press the [two Thunder] Offices onto your ears and let the breaths inside the ears produce the sounds that are the thunders (*wei lei* 為 雷). Your eyes flash three times, which is the lightning. Grind your teeth with one sound, which is the rumbling [of thunder].

In case that there is not any result, you just [force the points of pressure on your palms] to shift from *hai* (亥) to *ssu* (巳), and you must not clap open (*p'o-san*拍 散) [the *mudrâs*]. When the [desired] effects come about, you shift then [the hands with the *mudrâs* Thunder] Office to the position of your breast and clap open [your hands to give up the *mudrâs*]. You concentrate [your vision] on the Three Marshals (*san-shuai* 三 帥) and the Great Rumbling Divinity of the Five Thunders (*wu-lei p'i-li ta-shen* 五 雷 霹 靂 大 神), on the generals and emissaries of the *sun*-window (*sun-hu jiang-li* 巽 戶 將 吏) and the Thunder Divinities of the Eight Trigrams (*pa-kua lei-shen* 八 卦 雷 神). They force their way upwards to the window on the top of your head from where they leave. Thunder, lightning, rumbling, wind and fire join together and press forward as to mix with the heavenly thunder that was summoned, and they become one [unified] entity."

The religious side of Thunder magic deals with spirit generals and other divine charges that tend to have either personal or formalistic names. We sometimes find a combination of both elements.[64] Divine names may resemble common personal names that actually refer to deities of the Posterior Heaven (*hou-t'ien* 後 天). Such names often go with the divine ranks of meritorious Taoists who *post mortem* received divine Thunder ranks as a reward. The formalistic names most likely refer to abstract divine entities of the Anterior Heaven (*hsien-t'ien* 先 天). Thunder Magic unites them all.[65] It is a general conviction that the priest can summon and visualize all these divine entities as internal and cosmic realities.

Nevertheless, the sources of Thunder Magic frequently speak about spirit generals whom the priest may invite to descend (*chiang* 降) on himself and enter

---

64 See F.C.Reiter: „The Name of the Nameless and Thunder Magic", pp.97-116.
65 See F.C.Reiter: „The Discourse on the Thunders", p.222 sq.

(*ju* 入) some amulets. The spiritual and physical intermediation of the priest is instrumental to achieve this result. We may be tempted to speak of possession, which would claim some mediumistic or shaman features to matter.[66] On the other hand, we also know that divinities of the Posterior Heaven (*hou-t'ien* 後天) may potentially be present inside the human body where they were implanted when the practitioner received the respective registers. The registers constitute and specify the individual identity of the priest who in this way unites man-and-the-divine.

The compilation of ritual texts with the title Great Method of the Heavenly Lord Hsin, the Fierce Emissary who Carries the Wind on his Back (*Fu-feng meng-li Hsin t'ien-chün ta-fa* 負風猛吏辛天君大法) offers another telling description of how to activate the internal potentials and concentrate the divine forces (*lien-shen* 鍊神). In other words, we learn how the Taoist adopts a specific divine quality (*pien-shen* 變神). After the presentation of the pseudo-Sanskrit spell Departing from the Husk (*ch'u-ku chou* 出穀咒) we read the following words:[67]

Inhale deeply the purple breaths of the *sun*-direction (巽) and expel them sighing across the altar. You make then use of the two orifices (*ch'iao* 竅) below your tongue, and from the left side you emit chuckling the green [breath], from the right side the purple [breath], just the two breaths. They mix up and assemble to become (*chieh-ch'eng* 結成) the Grand Marshal (*ta-shuai* 大帥). Now, you recite the Written Oath (*shih-chang* 誓章) and lead [the breaths] into the Palace of the Gall (*tan-kung* 膽宮) where they assemble and become the Generalissimo (*yüan-shuai* 元帥) who has here his home palace.

The Written Address (*shih-chang* 誓章) documents the process of an internal spiritual promotion, which takes its speaker to be the divine agent. In other words, the swearing priest is in the position of his spirit *alter ego* who is the Generalissimo and Heavenly Lord Hsin (Hsin t'ien-chün 辛天君).[68] These are the words of the address:

---

66 Liu Chung-yü: *Tao-chiao fa-shu*, pp.55-60.
67 See TT 1220: 81.2a. At the end of chapter 81 we find the name of Pai Yü-ch'an 白玉蟾 (fl. 1st half of 13th ct.), which points to a rather late date of the Great Method in TT 1220 *Tao-fa hui-yüan*. This example is quite in accord with the expositions by Wang Wen-ch'ing (王文卿), which documents the continuity in the development of Thunder magic. Concerning the Fierce Emissary, see below the translation of chapter 56 of *A Corpus of Taoist Ritual*, TT 1220: 56.29a.
68 TT 1220: 81.2a-3a.

(2b) Fierce emissaries and divinities of Thunder and Thunderclaps, your might shakes the Nine Heavens and your thunderclaps are omnipresent in the Three Realms. You are loyal and diligent and support the god-emperors and lords (*ti-chün* 帝君) with your bodies that quickly rise up [to be as tall as] one million *chang* (丈). Your title of honour is Venerable of the Thunder Departments and Commander-in-Chief for All Thunder Departments (*lei-pu tsun tu-tu chu-lei pu* 雷部尊都督諸雷部).

Wind Earl, Master of the Rain, Mother of the Divine Sound of Thunderclaps and the Radiance of Lightning, legions of Night Demons of Grand Might on the left side and on the right side receive respectfully the heavenly order (*t'ien-ling* 天令) to assist and support the Venerable of the Five Thunders.

I have reverently received the decrees of the Jade Emperor (*yü-ti* 玉帝) and so can save the people in the world. In the case that there are people who have received and cultivated [the Taoist discipline],[69] I shall rapidly disclose to them my outer apparition. When [such people] ask me to ascend to the heavenly realms, I go to attend [the heavenly] audiences and submit [the petitions] to the god-emperors. When [such people] ask me to enter the Department of Earth (*ti-fu* 地府), I straightforward reach the Palace of the Dark Spheres (*Yu-ching kung* 幽境宮). When [such people] ask me to enter the Water Department (*shui-fu* 水府) the waves of the four seas then open up [a throughway for me]. When [such people] ask me to support the realm of life (*yang-chieh* 陽界), I establish the means to save all living beings. When [such people] ask me to save them in times of a drought, I let heavy rain come down. When [such people] ask me to arrest spirits and monsters (*ching-kuai* 精怪) I destroy and shatter their battalions. When [such people] ask me to help [a woman] in childbed, mother and offspring shall quickly separate their bodies and together with me live to the end of our lives, and we will jointly be servants of the Jade Emperor (*yü-ti* 玉帝).[70]

In the case that I, your minister (*ch'en* 臣), commit any offence against the heavenly laws (*t'ien-lü* 天律), my nine ancestors will suffer punishment by poison. **(3a)** In the case that I turn my back on you, the sun and the moon on

---

69 The formulation *shou-ch'ih che* (受持者) could be an abbreviation of *shou-lu hsing-ch'ih* (受籙行持), compare TT 1220: 69.11a-14a.
70 This is fact is a great survey on possible fields of operation for the ritual specialist of Thunder Magic.

the heaven above will dim and the wells and fountains will dry out in the earth down below. Grass and trees will no longer grow and for eternity I shall be a demon in the dark netherworld, not being able to ascend to pay a visit at [the Heaven of] Highest Purity (*shang-ch'ing* 上清).

Jen and *kuei* (壬 癸) mark the days when I descend. [The people] who have received and cultivated [the Taoist discipline] [71] must essentially be diligent and pure. There are the mandatory offerings of tea, jujubes and soup, and peach-wood incense must be burnt. I show up on the left side and on the right side with a peaceful heart that must not get startled, and together with you [72] I present the address to swear that I vow to save all living beings, all as ordered by the superior god-emperors. [73]

After the text of the oath, we find the following practical instructions. They ascertain that the speaker in the oath is the Generalissimo who is the spirit *alter ego* of the practitioner:

You circulate the black breaths of the *k'an*-palace (坎 宮) to let them pass upwards through the Jade Tower (*yü-lou* 玉 樓). You expel (*ho* 呵) then [the black] breaths in front of the altar and concentrate in meditation on the Marshal, who now mounts the ten thousand folds of black clouds. You [74] flash your eyes that send off the Generalissimo (*yüan-shuai* 元 帥) who eminently rises up. You form [the *mudrâs*] Thunder Office (*lei-chü* 雷 局) and together with one clap [of your hands] you shout out: Rapidly exorcise the urgent trouble! – In other words, the Generalissimo, "my *alter ego*", departs to fulfil the required ritual task.

The crucial point is the self-identification of the practitioner with the Thunder deity, the Generalissimo or Marshal Hsin (辛). He emerges out of the internal and bodily breaths and can be dispatched together with the black breaths. Sometimes it suffices to know the titles of the deities. This information is provided by the respective register (*lu* 籙) that the priest obtained from his teacher master. This is a most important aspect of Thunder rituals, which qualifies them as proper Taoist procedures. Taking into account the vast number of ritual methods (...*ta-fa* 大 法) that *A Corpus of Taoist Ritual* contains; we consider that they all must be genuinely connected with registers (*lug* 籙). On

---

71   ...and arranged the ritual...
72   In this case, "you" refers to the addressees of the oath who were named at the very beginning of the text.
73   Or translate: "...all in accordance with the law of ..." (一 如 上 帝 律 令).
74   The text says literally "the ritual master" (shih 師).

the other hand, we also remember that the patriarch Chang Yü-ch´u (張宇初) admonished his fellow Taoists to actualize (or respect) only one method, one register and one duty (*chih p´ei i-fa i-lu i-chih* 止 佩 一 法 一 籙 一 職).[75] The message is clear – the Taoist must not change his appearances and spiritual functions *ad libitum*, because he jeopardizes his religious identity.

There is still some another point that deserves our consideration. The deity has an outer form and appearance that the practitioner can identify. Wang Weng-ch´ing (王 文 卿) gave us his description of the divine apparition of Yen-huo Teng t´ien-chün (焱 火 鄧 天 君).[76] In the case of Marshal Hsin (辛) the description is a bit more modest. We learn that Marshal Hsin (辛) "wears an ox-ear cap,[77] has red hair, an iron face and silver teeth like daggers [that sharp]. He is clad in green cloud-fur-garments (*yün-ch´iu* 雲 裘) and black boots. His left hand holds the Thunder files (*lei-pu* 雷 簿). His right hand holds the Thunder brush (*lei-pi* 雷 筆). An abundant radiance of fire is seen above his apparition".[78]

Such marks of identity and status symbols (*hsiang-hao p´in* 相 好 品) help the Taoist in meditation and during rituals to link an apparition with a specific name that should accord with the register that he holds. This conviction is well established in Taoism, which much older canonical encyclopaedias and other descriptive texts document, for example, *San-tung chu-nang* (三 洞 珠 囊) and *T´ai-shang Lao-chün chung-ching* (太 上 老 君 中 經).[79]

We may ask who actually is entitled to set to use the ritual program of Generalissimo Hsin (Hsin yüan-shuai 辛 元 帥). Any Taoist is entitled to do so if he holds the appropriate register (*lu* 籙) of the Marshal with the names of the divinities that are involved. I have already indicated that this is the precondition

---

75 See TT 1232 *Tao-men shih-kuei* 11b. Concerning this text see, F.C. Reiter, in: *Companion* p.975.
76 See above. See the following translation of *chapter 56* of *A Corpus of Taoist Ritual*, TT 1220: 56.29a.
77 See H.A.Giles: A Chinese-English Dictionary, nr.9404.
78 See TT 1220: 81.1a, the introduction to the Marshal Class (*shuai-pan* 帥 班).
79 See TT 1168 *T´ai-shang Lao-chün chung-ching*, see, K.Schipper, pp.92-94 in: *Companion*. K.Schipper tentatively dates the text to the Later Han Period, specifying his earlier assessment, following Ch´en Kuo-fu and in comparison with his "Le Calendrier de Jade, Note sur le Laozi zhongjing", in: *Nachrichten der Gesellschaft für Natur-und Völkerkunde Ostasiens* 125, pp.75-80. As to *San-tung chu-nang* (三 洞 珠 囊), see F.C.Reiter: Der Perlenbeutel aus den Drei Höhlen, pp.129-157 (concerning TT 1139 *San-tung chu-nang* 8.1a-24a).

for the realisation of all the great methods or rituals (*ta-fa* 大 法) in *A Corpus of Taoist Ritual*. In the case of Generalissimo Hsin, the performing priest had received such registers (*lu* 籙). The spiritual might and presence of the spirit Generalissimo thus was implanted into the spiritual household of the priest. In other words, the priest who just addressed the deity with his "written oath" was able to activate this divine potential and adopt the identity of the Marshal/Generalissimo Hsin (辛 帥). He is also in command of the subordinate divine charges, whose might derives from their cosmic dimension that their titles reveal:[80]

(1a) The Savage Thunder Emissary of the East, Ma Yü-lin (*Tung-fang man-lei shih-che* 東 方 蠻 雷 使 者 馬 鬱 林);

The Savage Thunder Emissary of the South, Kuo Yüan-ching (*Nan-fang man-lei shih-che* 南 方 蠻 雷 使 者 郭 元 京);

(1b) The Savage Thunder Emissary of the West, Fang Chung-kao (*Hsi-fang man-lei shih-che* 西 方 蠻 雷 使 者 方 仲 高);

The Savage Thunder Emissary of the North, Teng Kung-ch´en (*Pei-fang man-lei shih-che* 北 方 蠻 雷 使 者 鄧 拱 辰);

The Savage Thunder Emissary of the Centre, T´ien Yüan-tsung (*Chung-yang man-lei shih-che* 中 央 蠻 雷 使 者 田 元 宗)

We see that Sung Thunder Magic unites a vast variety of Taoist concepts that develop in a new context. The actual presence of divine forces that have names and titles is an important aspect, which points to Thunder deities that either may descend from heaven or become apprehensible in a meditative visualisation that is based on the internal energies. This effort results in the spiritual formation of deities who actually become the spiritually acting agents in a ritual.

The writing of Thunder amulets is a great case in point. Wang Wen-ch´ing (王 文 卿) presents a rather detailed didactic explanation about the writing of amulets. And again, we study the didactic dialogues of Wang Wen-ch´ing (王 文 卿) and his disciple Yüan Wu-chieh (袁 無 介). We also study the subsequent commentaries by Yü-feng (御 風).

---

80 TT 1220: 81.1a-1b.

The Basis for Writing Amulets (shu-fu 書符)

The biography of Yeh Ch´ien-shao (葉千韶) in the introduction to this book already made mention of amulets. Heavenly amulets were a major element of initiation for the priest and exorcist Wu Meng (吳猛).[81] Amulets are certainly outstanding and often eye-catching expressions of Thunder Magic. The collection *A Corpus of Taoist Ritual* provides fascinating evidence. Wang Wen-ch´ing (王文卿) offers a very telling discussion of this theme:[82]

(14a) Yüan Wu-chieh (袁無介) sets forth his question concerning the secret instructions for the writing of amulets. He wishes to learn something about this mystery.

[Wang] Shih-ch´en ([王]侍宸) makes [the following] statement: When the one breath is present, it can be used above to reach the heavenly perfected ones, and down [on earth] it can be used to subdue the bewitching demons (*yao-mei* 妖魅). In the middle [the one breath] can be used to arouse and agitate wind and rain, thunder and lightning. At the time when an amulet has to be written you must first fix your breathing (*hsi* 息),[83] holding the writing brush [in your hand]. You lead on the pure breath with your nostrils, extending for long the one action of inhaling. [The breath] must not be turbid. It is most important that the breath is pure. Then, you shut up the breath and hold on to it without any breathing at all (*hu-hsi* 呼吸). You use speedily the writing brush to write down the amulet. Having done [the writing] you let the splendour of your heavenly eye (*t´ien-mu* 天目)[84] enter [the amulet], and all the generals and emissaries (*chiang-li* 將吏) that you summoned enter the centre of the amulet. Your mouth releases strongly coughing the breath right above the amulet, and golden radiance will cover [the amulet]. In your meditative vision you see the generals and emissaries who were summoned and are [now] inside the amulet that you wrote. **(14b)** You speedily take the three characters "vast", "clear" and "bright" (*hung* 泓 *ch´eng* 澄

---

81 See my Introduction.
82 TT 1220: 69.14a-16a. This text is a major source in my article "The Management of Nature: Convictions and Means in Daoist Thunder Magic (Daojiao leifa)", pp.193-210, in: AAS 29. The writing of amulets is a notorious theme in almost any compilation concerning specific Thunder rituals. They all involve similar elements; compare for example TT 1220: 219. 20a-20b (*shu-fu nei-pi* 書符內秘) in: *Shen-hsiao tuan-wen ta-fa* (神霄斷瘟大法). Generally see C.Despeux: "Talismans and Sacred Diagrams", pp.498-540, in: L.Kohn ed.: Daoism Handbook. C.Despeux gives a general survey on the variety of amulets and graphic representations of spiritual potentials and elements.
83 This means the inhaling and the exhaling of breath.
84 For this term, see above p. 11b, and TT 1220 *Tao-fa hui-yüan* 56.31a.

ming 明) [85] and [mentally] stamp (*ya* 押) them onto the amulet. If the breath in your mouth leaks out, during the time when the amulet is being written, or if your mouth does not enclose the breath and an amulet is still being written, such an amulet does not have any divine force. Even if there should eventually be some divine result, this occurs [only] by chance once a time. Such an effect is not due to the force of the amulet.

The commentary by Yü-feng (御 風) says: Amulets do not have an orthodox outer form (*cheng-hsing* 正 形). They have their divine force (*ling* 靈) based on the breath (*ch'i* 氣). The divine force (*ling* 靈) is the ancestral breath (*tsu-ch'i* 祖 氣). In case that the ancestral breath is not clearly present (*ming* 明), how can you still wait for its divine force (*ling* 靈) to become effective? When the divine force is present, you know where the ancestral breath stays at and when it comes and goes. You firmly guard the cinnabar fields (*tan-t'ien* 丹 天) [86] and gently nourish that basis [of your body]. Generally, when the human hands, feet, eyes and ears, nostrils and tongue set out to move, [the same time] these orifices (*ch'iao* 竅) are also agitated. [87] Concerning these orifices, they are the locations that assemble and contain the original breath. They are the bags that collect and store the essences and spiritual forces. The *hun*- 魂 and *p'o*- 魄 souls are guarded at the purple window (*tzu-hu* 紫 戶). The purple window is the gate of fate (*ming-men* 命 門) and the location that connects [the embryo] with the womb. The three [cinnabar fields] must not desert each other. [88] This is what *Meng-tzu* (孟 子) calls the breath that rears the natural greatness. [89]

In any case, when amulets have to be written, you purify your mind, **(15a)** tranquilize your considerations, congeal your spiritual forces, fix your breathing and grasp the writing brush. You use the nostrils to lead on the pure breath and having completely inhaled it you let [the pure breath] return to the central bar (*chung-kung* 中 扃) [90] that you firmly shut. You let the breath of pure-*yang* (ch'ing

---

85   The the radical rain (yü 雨) tops the three characters. They are ritual devices that we do not find in official dictionaries. The pronunciation derives from the basic characters as indicated. They frequently appear in print on modern amulets (Taiwan). Here, however, the priest most likely deposits them mentally on the amulets.
86   TT 1250: 15b
87   This word is usually translated "orifice", but here *ch'iao* (竅) should refer to the cinnabar fields. Compare H.A.Giles: A Chinese-English Dictionary, nr.1426.
88   This points most likely to the interdependence of the three cinnabar fields.
89   *Hao-jan chih ch'i* (浩 然 之 氣); see H.A.Giles: A Chinese-English Dictionary, nr. 3891 ("passion-nature").
90   The "central bar" seems to be identical with the "yellow court", see below the Commentary by Yü-feng (御 風).

yang chih ch'i 清陽之氣) amass and flow into the central palace (chung-kung 中宮). Then, you shift to open the central bar, and the original breath (yüan-ch'i 元氣) comes forth. You blow it onto the tip of your writing brush, and so a golden radiance flashes brightly with all its splendour. You write the amulet on the paper, which is like the force of a dragon that crawls on. All the paper is filled with the killing breaths (sha-ch'i 煞氣) that form unseen barricades. Again, your nostrils breathe in pure breath. You let [the pure breath] straightforwardly pass through the heavenly joints (hsüan-kuan 玄關). The nine orifices above and below must all be shut up, and they must not allow any leakage to happen.[91] The one breath completes the amulet. When [the writing of the] amulet is completed you release the breath and blow it onto the amulet. You take the three characters hung, ch'eng and ming and stamp them [mentally] on the amulet. Now, when the amulet is being written you must not let the original breath (yüan-ch'i 元氣) leak out, because otherwise the amulet will not be magically efficient (yen 驗).

[Wang Wen-ch'ing (王文卿) continues saying:] At the moment you are writing an amulet you leak the breath from your mouth. You should a second time enclose the breath and [again] write down this amulet. In case that you leak the breath three times and the writing of the amulet is not identical at all (pu-i 不一), then you have the result that the heavenly generals do not respond and what had been done ritually will not be magically efficient. Generally, the essential [requirement] for writing amulets is the absolute **(15b)** enclosure of the one load of breath (pi-ch'i 閉氣) in the mouth. When one encloses this breath, the individual mind (wu-hsin 吾心)[92] must not have any erratic thoughts. When the own will is concentrated, the source of perfection is thus secure and firm. When you blow [the breath] upon the amulet, you lend it (chieh 借) the divine might of the one breath. In case that you leak away the breath, what else is still at hand to be used? Those who practise rituals today do not know [the method concerning] the orifices, and it is for this reason that all their writing of amulets is ridiculous.

The commentary by Yü-feng (御風) says: Concerning the time when an amulet is being written, it is all about enclosing the breath (pi-ch'i 閉氣). It is essential that the nostrils first collect the pure breath that you transfer then to enter the yellow court (huang-t'ing 黃庭). You shut up the nine orifices and achieve that

---

91 Concerning the nine orifices (ch'iao 竅) in a context with the hun- and p'o-souls (魂魄), see below and compare TT 110 Huang-ti yin-fu ching shu 1.6b-7a.
92 Literally: „my mind"

the original breath is complete and present. You tip down the brush and write the amulet. One [stroke of the] brush sweeps and completes the golden radiance that flares up. If you perform in this way, you will get an echo and a response. In case that you leak the original breath and also do not use the nostrils to lead on the pure breath to be forcefully stimulating, how could you let it happen that the original breath comes forth? Anyway, when you enclose the breath your mind then does not shift, move, and harbour any mixed thoughts. Your whole will rely on the very centre. Breath and brush all revolve [together], and in this way your will definitely completes the amulet. Gentleman who practise rituals **(16a)** [but] do not attain the transmission [concerning] the heavenly joints (*hsüan-kuan* 玄 關) and [furthermore] are not informed about the ancestral breath, practise in vain the writing of amulets and deserve to be greatly ridiculed."

The text explains the internal involvement and the physical requirements for writing amulets. We understand that divine forces inhabit the amulet and its graphic design, which therefore carry divine potentials. We do not need to consider any theories about the value and meaning of symbols. We simply learn and accept that there are spirit generals and their entourage of spirit emissaries who populate the amulet. The Taoist has the privilege to visualize a transcendent reality that escapes the apprehension of a layperson. In short, the amulet is not so much a combination of graphic designs on paper, but it is an energetic and spiritual agent. Small wonder that some amulets may be drawn in the air, in empty space, and there is not any graphic trace at all.

We may well ask what can be done about amulets that are not magic and do not show any advantage for its owner. Many amulets, for example, were used to procure rain. Now, what do we do if it was all in vain?

The short tract The Soaring Sword Beheads the Heavenly Emperor (*fei-chien chan t'ien-huang* 飛 劍 斬 天 皇) is a good example. The text says that after two or three days when the amulet failed to procure rain, the priest could use the following method: [93]

(16a) You take one leaf of strong yellow paper and write on it a Heavenly-Emperor amulet (*t'ien-huang fu* 天 皇 符). The head of it must point to the direction of Southeast (*sun-fang* 巽 方) and the feet [of the amulet] to [the direction] of Northwest (*ch'ien-[fang]* 乾 [方]). In the middle of the altar and below the flayers, at night around the time 11 p.m. until 1 a.m. (*tzu-shih* 子 時),

---

93 TT 1220: 68.16a-16b (*Lei-t'ing san-ssu ch'i-tao pi-chüeh* 雷 霆 三 司 祈 禱 秘 法).

you take the blood of a cock and spit it onto the Heavenly-Emperor amulet (*t'ien-huang fu* 天皇符). The next day around noon (*wu-shih* 午時) when the sun stands in its zenith, you [94] form the *mudrâ* Thunder Office (*lei-chü* 雷局) with your left hand and your right hand holds the sword. Standing vis-à-vis the Heavenly- Emperor [amulet], you visualise the Heavenly Emperor with his human head and the body of a snake. [95] [You visualize] the waterwheel on the top of his head, and his feet tread on a waterwheel. [The Emperor] raises rapidly his body to be as tall as ten thousand *chang* (丈), and your own body (*wu-shen* 吾身) also rises to be as tall as ten thousand *chang* (丈), being in a great rage. You visualise that the heaven dims and the earth gets black [and dark]. In one moment, you speedily turn around your body and using your sword cut the Heavenly-Emperor amulet into **(16b)** two pieces. This is called the Soaring Sword Beheads the Heavenly Emperor (*fei-chien chan t'ien-huang* 飛劍斬天皇).

In other words, the Taoist takes on the tall and awful apparition of the Heavenly Emperor himself, and having the same spiritual capacity he destroys the amulet that certainly did not house the Heavenly Emperor.

The text School Talks gives another informative report that may serve as a good documentation concerning amulets. It features the failure of the application of an amulet to alleviate the delivery of a child. We also learn why this effort failed:[96]
(4a) …Being once in a private home [I, your] disciple administered an amulet to save a woman who was about to deliver a child. When I summoned the [spirit] generals, my heart was greatly shaking and my left eye repeatedly flickered. [97] When I issued the amulet and dispatched the emissaries, my heart and my will were not pleased, and the Emissary of the Will (*i-shih che* 意使者) had not yet arrived at the scene. It worked just like that. The next morning the family bade me farewell and told me that the prospective mother had visualised a demon emissary with a green face and three eyes. **(4b)** Fire charged forth from the eyes of the demon engulfing the body of the woman in childbed, and she felt that her

---

94 It says litarally: "the ritual master (fa-shih 法師)".
95 See, for example TT 1220: 67.10b, and 124.12a, which show the importance of the image of the snake, eventually representing the lightning in Thunder Magic. Also see TT 1220: 69.24a, for a commentary by Yü-feng (御風), who associates the snake (she 蛇) with lightning, fire and the trigram *li* (離). See F.C.Reiter: "A Preliminary Study of the Taoist Wang Wen-ch'ing (1093-1153)", in: ZDMG 152, pp.169-170.
96 TT 1250: 4a-4b. The participants in the conversation are Wang Wen-ch'ing (王文卿) and Yüan T'ing-chih (袁庭植).
97 Lit. "jump" (t'iao 跳).

body became as hot as fire. After a short time, she delivered a dead born baby. Please, what is the meaning of this event?

[Wang Wen-ch'ing 王文卿] gives the following answer: This is a very clear result. How do you come to be that stupid? There was the great shaking in your heart, which means that the Emissary of Your Ancestral Palace (*ju tsu-kung shih che* 汝祖宮使者) was immediately transferred to your heart as soon as he had appeared, and this is why your heart was excited (*chi* 急). As to the green face and the three eyes, this was your own divine force (*ju shen* 汝神) that following the amulet underwent mutation and change. Your left eye flickered, this means that [the element] fire returned to wood-county (*mu-hsiang* 木鄉). The mother who under this condition gave birth to [a child] had to lose the child. This principle is clear. What need is there at all to have that many doubts and questions?"

The conversation seems to show that the disciple did not properly control his internal processes in a way that would have transferred to the amulet the required positive force. Obviously, the dispatch of the spirit Emissary the Will did not work. The vital wood-energies should have been transferred to the womb. However, the Taoist in vain consumed them with his beating heart and flickering eye. Anyway, we understand that the dispatch of spirit emissaries and generals goes with amulets, which seems to be a very demanding and responsible effort.

We know that all of this is but a small glimpse of the antique and yet steadily present traditions concerning the application of amulets. Wang Wen-ch'ing (王文卿) and his fellow Thunder specialists participate in this stream of traditions that a text with the title On Writing Amulets (*shu-fu shih* 書符式) in the collection Great Shang-ch'ing and Ling-pao Rites (*Shang-ch'ing ling-pao ta-fa* 上清靈寶大法) testifies:[98]

(2a) The ritual method says: In case that the time for writing an amulet has come, you face the East in a pure room (*ching-shih* 淨室). Pay your respects [to the divinities] and kneel down for long. Visualize your own body to be [the Heavenly Worthy] of Prime Origin (*yüan-shih* 元始).[99] There are 10.000

---

98   On this text, see J.Lagerwey: *Companion* pp.1021-1024. This corpus of Taoist texts was compiled by Wang Ch'i-chen (王契真); transmitted by Ning Ch'üan-chen 寧全真, 1101-1181). As to the translation, see TT 1221 *Shang-ch'ing ling-pao ta-fa* 14.2a-2b.

99   For another example, see TT 1221: 40.19b.

divinities in attendance all around. Your body has a radiant shining. You breathe in the breaths in the five colours of the five directions, and your hands twist firmly the *mudrâ* Jade Purity (*yü-ch'ing chüeh* 玉清訣). You use your nose to lead on the nine breaths to enter your mouth and swallow them nine times. You circulate the breaths in the cinnabar fields to ascend and push against the Mud Pill [Palace] (*ni-wan [kung]* 泥丸 [宮]). The breaths flow throughout the whole body. They evenly spread in all the joints [100] **(2b)** and release a great radiance that illuminates all around. You see (*chien* 見) all the heavens, the earth, the sun, the moon and the constellations. After this, you recite the secret spells twelve times, collect the breaths and lead them into the writing brush. You get up your body, pay again your respects [to the divinities], grasp the writing brush and grind your teeth thirty two times. You face then the East in order to write out the amulet. For its upper part, you take the stars and the dipper to top the structure [of the amulet]. You concentrate on the astral constellations, on sun and moon and all their radiance that links up with the breaths of the [secret spells] in Five Paragraphs (*wu-p'ien* 五篇) [101] and illuminates the inside and the outside [of your body]. You collect and treasure [these breaths]. When the time of their application has come, you should visit the Gate of Heaven (*t'ien-men* 天門), burn incense, again pay your respects and kneel down for long. You silently memorialize to the [Heavenly Worthy] of Prime Origin and clearly state all the matters of concern. After that you can apply [this method], and there will always be the [appropriate] response. Keep the method secret and treasure it dearly."

This text nicely summarizes the efforts that may lead to the composition of an amulet. The amulet appears to be a cosmic entity just like the acting priest himself who assumes the cosmic dimension of Prime Origin. Certainly, this is most remarkable. [102]

Praying for Rain (tao-yü 禱雨)

A most spectacular and important application of Thunder Magic is certainly the ritual effort to procure rainfall. We must not forget that Taoists lived in an agrarian country, and harvests depended on fine weather conditions. In the case

---

100 or "orifices"
101 This refers most certainly to the following paragraph "writing five paragraphs of secret spells" (*shu wu-p'ien mi-chou* 書五篇密咒), TT 1221: 14.2b-3a.
102 For a modern and very conclusive description of Taoist amulets and the related practices as understood today, see Chang Chih-hsiung (張智雄) and Li Feng-mao (李豐楙): *Cheng-i fa-fu yü tao-chiao wen-hua* (正一法符與道教文化).

of droughts, people asked for the help of the Taoist specialist. They believed that the specialist in Thunder Magic rituals would have the means to cope with both, droughts and inundations. Wang Wen-ch'ing (王文卿) himself was famous for his ability to handle weather conditions. He allegedly was successful when he prayed for a clear sky in support of the imperial sacrifices that emperor Sung Hui-tsung (宋徽宗) wished to perform at the Bright Hall (ming-t'ang 明堂). We find a report about this event in the canonical biography of the Taoist.[103]

Wang Wen-ch'ing (王文卿) gives lucid explanations about the spiritual and ritual efforts that are required to procure rainfall. The didactic chapter with the title Praying for Rain (tao-yü 禱雨)[104] shows the application of formal and cosmographic elements, which, for example, the eight trigrams (pa-kua 八卦) represent. We certainly understand rain and drought to be natural or even cosmic phenomena that are well beyond the confines of the human body. Now, we explicitly learn that "thunder, lightning, rain and hail are all within the own body (wu-shen 吾身), in fact in the very centre of the two kidneys. [Thunder, lightning, rain and hail] come forth [from there] uniting with the breaths of heaven and earth. This is based on the correctness (cheng-chih 正直) of heaven and earth, both being impartial. Man, on the other hand, can grasp the creative forces of heaven and earth. The cosmos is within his hands and all mutations are born out of the human body".[105] I now present the tract Praying for Rain (tao-yü 禱雨):
(17b) Yüan Wu-chieh (袁無介) stets forth his questions about the definition of the essentials for the prayers for rain.

[Wang Wen-ch'ing (王文卿)] gives the following answer: As to the important instructions concerning the prayers for rain, it is just like the writing of amulets and the dispatch of [spirit] generals. I must sit cross-legged either at the altar or in a cell [for meditation], **(18a)** harmonise (t'iao-hsi 調息) my breathing and visualize in meditation my heart that is like a lotus flower that has not yet

---

103 A translation of the biography gives: F.C.Reiter: "A Preliminary Study of the Taoist Wang Wen-ch'ing (1093-1153)", in: ZDMG 152, pp.160-169, see esp. p.166 (TT 296 *Li-shih chen-hsien t'i-tao t'ung-chien* 53.16a-21b)
104 See TT 1220: 69.17b-21a
105 See TT 1220: 69.24a. The last two sentences in this quotation were taken from the *Huang-ti yin-fu ching* (黃帝陰符經), compare TT 110 *Huang-ti yin-fu ching shu* 1.1b. Concerning the *Huang-ti yin-fu ching* (黃帝陰符經) see, F.C.Reiter: "The Scripture of the Hidden Contracts (*Yin-fu ching*): a short survey on facts and findings", in: Nachrichten der Gesellschaft für Natur- und Völkerkunde Ostasiens 136, pp.75-83 (1984).

opened its blossom. There is red breath that moves straight downwards to the space between the two kidneys.[106] I behold the vision of a clear pool of perfect water (*chen-shui* 真 水) that is right between the two kidneys. I [now] ponder on how the red breath in my heart descends downwards, whereas the [perfect] water streams bubbling upwards, firmly encloses the red breath and makes its way from the liver to exit (*ch'u* 出) at the root of the tongue. I only visualise the cloudy breath (*yün-ch'i* 雲 氣) in my mouth that suddenly exits and in front of my face revolves, passes the *sun*-window (*sun-hu* 巽 戶) and gradually becomes as big as the wheel of a carriage that rolls on and ascends to heaven. Those clouds fill overflowing the cosmos (*liu-ho* 六 合). On the side of the ears, there is the sound of wind and thunder that rumbling become clearly discernible. I fix then my breathing and expel coughing the breaths nine times. It is just the same as in my preceding [expositions on] the internal practice (*hsing-ch'ih* 行 持).[107] [Expelling the breaths] nine times I finally feel that the water of the kidneys in my body really ascended. The need to urinate (*hsiao-i* 小 遺) becomes urgent. I absolutely must not give in. In case that I go to find relief and doing so ooze out the water from the kidneys, the rain is not going to fall. I straightaway keep on waiting for the wind and the rain to come to [the area of] the altar, and to be substantial. After this has happened I rise up and slowly urinate, but the heavy rain has then already come. It is only this internal practice, which never fails. Is it agreeable **(18b)** not to be careful in this matter?

The commentary by Yü-feng (御 風) says: The important instructions concerning the prayers for rain are a top mystery (*chih-miao* 至 妙). The principles of the great mystery lay in the two kidneys. The water of the two kidneys is procured (*sheng* 生) through the trigram *k'an* (坎).[108] *K'an* then has [in it] pure *yin* (陰) that is the straight symbol of the North. Its straight symbol *k'un* (坤) [stands for] the [number] Six and [the element] earth (*liu-t'u* 六 土).

---

106   The statements in our texts concerning the intestines do not reflect the older traditions of TT 432 *Huang-t'ing nei-ching wu-tsang liu-fu pu-hsieh t'u* (9th ct.).
107   Compare TT 1220: 69.11a-14a, No.3 Internal Practice (*hsing-ch'ih* 行 持), especially pp.11a-11b. The wording of these two paragraphs is very much the same.
108   Compare TT 263 *Hsiu-chen shih-shu chin-tan ta-ch'eng chi* 10.12b (*wen nei-wai pa-kua* 問 內 外 八 卦), which seems to show that the commentary by Yü-feng (御 風) is a rather late text. The citations of this text (TT 263) remind us that the terminology of Thunder Magic reaches beyond any specific Taoist school. Actually, there is no great point in pinning down all the statements of Thunder Magic in TT 263 and similar late sources. The starting point for these statements is the notion of *water* and more specific, the water from the two kidneys. *K'an* (坎) has two *yin*-strokes only whereas *k'un* (坤) consists of three *yin*-strokes, see *Chou-i fan-lieh* p.9, in: *Chou-i*, ed. Kanbun taikei Tokyo 1913.

Does it not well enclose *ch'ien* and [the element] metal (乾 金)? Metal gives birth to water. Water has [the potential of] life of heaven and the One (*t'ien-i* 天 一). Heaven and the One are connected with the River Chart (*ho-t'u* 河 圖). The One and the Six lodge at *k'an* (坎). The [number] One stands for the western region and the region of pure metal. The [number] Four (*ssu* 四) gives birth to [the element] metal. The [number] Nine of [the element] metal is *yang* (陽), and the final stage is [reached] when *yang* (陽) at its zenith gives birth to *yin* (陰). *Yin* (陰) is *k'un* (坤), the [number] Six and the [element] earth (*liu-t'u* 六 土). Does it not well enclose *ch'ien* and the [element] metal (乾 金)? The [element] metal is *yang* (陽), and *yang* (陽) is the dragon. The dragon (*lung* 龍) has a pool (*t'an* 潭), and the pool holds the water and the rain. Rain emerges from the breath of earth, and the clouds emerge from the breath of heaven. When the clouds steam, then it is that rainfalls occur. The nine strokes [of the writing brush] for the character *yü* (雨), "the rain", are connected with heaven, and this is the inherent mystery of *yang* (陽).

In the thunder department (*lei-pu* 雷 部) there are the wind and the clouds, thunder and lightning, mist and hail, snow and rain **(19a)**. These eight pure matters borrow the Nine, the inherent number of [the element] metal of the River Chart. When it is that *yang* (陽) finds its final stage, the ninth matter then is the rain. [109] Rain emerges from the breath of the earth. The earth (*ti* 地) is connected with [the trigram] *k'un* (坤), and *k'un* stands for [the element] earth (*t'u* 土) that dominates *yin* (陰), and *yin* (陰) [in turn] procures (*sheng* 生) the rain. [110]

You first concentrate [your internal vision] on the red lotus blossom in your heart. The *yang*-breath (陽 氣) steams and descends to enter the one clear pool of water between the two kidneys. The one stroke in the middle of [the trigram] *k'an* (坎) is the minor-*yang* (*shao-yang* 少 陽). When the minor-*yang* is born (*sheng* 生) for the first time, it is stored within [the element] earth (*t'u* 土), and then the *yang*-breath (陽 氣) within [the element] earth gradually expands and exceeds. Generally, when the heavenly time has come to let it rain, the *yang* (陽) steams and [consequently] there is rainfall. The stones become shining (*jun*潤) and emit water. When water comes forth [from the stones, we know that] the indicator of

---

109 The preceding list named nine matters, the last one being "rain". In other words, the preceding eight phenomena build up *yang* (陽) that culminates in [the number] Nine and in the natural phenomenon that is the rain.
110 Compare J.Blofeld: I Ching, The Chinese Book of Change, pp.216-217. London 1976.

rain has appeared. *Yin* (陰) encloses the *yang-p'o* [-souls] (陽 魄) and the chill of the nine heavens (*chiu-hsiao han* 九 霄 寒).

You collect first the breath of complete *yin* (陰) that is contained within [the trigram] *k'an* (坎). You let the breath [of complete *yin* (陰)] ascend to the double pass of the spine and reach the windlass pass (*lu-lu ta-kuan* 轆 轤 關) where it enters as far as three *ts'un* deep into the window of the brain (*nao-hu* 腦 戶), and [the location] there is called Bright Hall (*ming-t'ang* 明 堂). [This procedure] is tantamount to the installation of the thunder altar (*lei-t'an* 雷 壇) on the top of [Mount] K'un-lun (崑 崙).

A second time you let ascend (*sheng* 陞) the breath of red-*yang* (*hung-yang* 紅 陽) that is inside the *li* [-trigram] (離)[111]. [Trigram] *li* (離) has two *ch'ien* (乾) [-strokes] that enclose one *k'un* (坤) [-stroke].[112] [The element] earth (*t'u* 土) is located **(19b)** inside. Pure *yang* (*ch'un-yang* 純 陽) is on the outside, and pure *yin* (*ch'un-yin* 純 陰) is on the inside.[113] *Yin* (陰) is water.[114] There is the saying: The Dipper of the South (*nan-tou* 南斗) moulds the *hun*- [souls] (魂). It is the water, which actually moulds.

The womb of [the element] water (*shui-t'ai* 水 胎) is set up at the position of the trigram *li* (離). *Li* (離) is the sun. There is a black radiance within the sun, and this [black radiance] exactly is the water. The essence of *yin* (陰) is within *yang* (陽), and so the three worlds (*san-chieh* 三 界) are clear and bright, which belongs to the theories for the prayers to attain a clear sky.

In the case of prayers for rain you must "boil the mountains and cook the sea" (*p'eng-shan chu-hai* 烹 山 煮 海)[115] to get rain. You just let the breath of red-*yang*

---

111 This is a name for the heart.
112 *ch'ien* (乾) and *k'un* (坤) represent *yang* (陽) and *yin* (陰).
113 This refers to the trigram *li* (離). Compare *Book of Changes*, p.li (*Introduction*).
114 Compare Hsü Chien (徐 堅): *Ch'u-hsüeh chi* (初 學 記) 6 (*tsung-tsai shui* 總 載 水) p.111, quoting *Huai-nan tzu* (淮 南 子) and *Han-shu* (漢 書). Ed. Peking 1980.
115 The phrase describes some meditative practices and the appropriate symptoms, see TT 1220: 85.6b "first call upon the fire of the heart. Boil the water of the kidneys, and seemingly within a moment there is a painful heat within the belly, and that exactly is the cooking sea that spills over. Following this, you draw the water of the kidneys as to enter the top [of the head], the boiling mountain. Immediately your forehead transpires cold sweat. This means that the boiling mountain moves". This is said to be a secret internal method. However, in our text the correspondences and associations are different. This is a nice example for an emblematic phrase that everybody re-thinks or re-interprets following the own requirements.

(*hung-yang* 紅陽) ascend to reach the double pass of the spine, and then let it enter three *ts'un* deep into the window of the brain (*nao-hu* 腦戶). This is called the Appointment at the Bright Hall (*ming-t'ang chu-cha* 明堂駐劄).

You first grasp the *yin* (陰) of the *k'an* (坎) [-trigram]. Overflowing black breath covers and rests upon the top of your head. Heaven blurs and bends down on the pagoda, and the four mountains stand amidst blurring mist with drizzling rain [all around]. You let the breath of red-*yang* (*hung-yang* 紅陽) ascend and directly collide with[116] the breath of *yin* (*yin-ch'i* 陰氣). When you see that the red breath arrives, you only see an abundant steaming. Again, you let ascend the *yang* of *k'an* (*k'an-yang* 坎陽) that is a perfectly black breath, which directly collides with the top of your head (*ting-hsin* 頂心). Two *yin* [-strokes of *k'an* (坎)] enclose one *yang* [-stroke] and, definitely, they produce the abundant steaming. Then it is that the stones moisten and emit water.

When a brilliant and steaming blaze of fire occurs in the middle of the sun, the fish tail then turns over the waves (*yü-wei fan-p'o* 魚尾翻波), the torpid insects (20a) struggle to get out of the window of earth (*ti-hu* 地戶) and the sound of thunder certainly shakes the night. When this happens, the radiance of lightning blazes,[117] in the North black pigs (*wu-chu* 烏豬) cross the River[118] and black clouds cover the dipper (*chao-tou* 罩斗), [and so we know that] the indicators of rain have appeared.

When in the morning black breath ascends and black clouds cover the base (*chiao* 腳) [of the land] late in the evening, heavy rain will certainly come. The subtlety [of these instructions] is contained in the basic scriptures that the teacher-master transmits.[119]

You focus [your meditation] on your heart that resembles a lotus blossom that has not yet opened. [You visualize] the red breath that is inside and sinks straight down to the two kidneys. You see the one clear pool of perfect water (*chen-shui* 真水). The red breath of the heart sinks down, and this [perfect] water bubbles up enclosing the red breath. [Starting out] from the liver [the red breath]

---

116 *ch'ung* (衝) "collide with" or alternatively "forces its way to"
117 I suppose that *shan* (扇) stands for *shan* (煽), compare H.A.Giles: A Chinese-English Dictionary nrs. 9669,9672.
118 The word River points to the Milky Way.
119 Actually, we cannot identify so far central scriptures in the Thunder Magic of that time. Taking the genre of this text into account, the teacher-master should be Wang Wen-ch'ing (王文卿).

passes the heart and exits from the base of the tongue. You only feel the cloudy breath (*yün-ch'i* 雲氣) in your mouth (*wu-k'ou* 吾口) [120] that suddenly exits and in front of your face revolves, passes the *sun*-window (*sun-hu* 巽戶) and gradually becomes as big as the wheel of a carriage to roll on. These clouds fill overflowing the cosmos (*liu-ho* 六合).

The commentary of Yü-feng (御風) says: [121] Clouds are *k'un* (坤) and [belong to the element] earth (*k'un t'u* 坤土). The element earth dominates [the colour] black (*hei* 黑). However, *k'un* and earth (*k'un t'u* 坤土) are enclosed inside the trigram *li* (離). When *yang* (陽) reaches its zenith, it procures *yin* (陰) and *yin* (陰) is black. The colour black stands for [the elements] water and earth. **(20b)** The wind comes to life at the centre of [the trigram] *li* (離). When the wind rises, the fire burns. Heaven and [the number] Three give birth to [the element] wood that is stored in the trigram *li* (*li-kua* 離卦). *Li* (離) lodges at the position of the East. The third transmission after "heaven" procures [the element] wood and is called [trigram] *chen* (震). It is the eldest son (*chang-nan* 長男).[122] [The element] wood is hidden inside the trigram *li* (*li-kua* 離卦). [The element] wood procures fire, wind and thunder. They are thus all born within [the trigram] *li* (離). Thunder and rain come to life on the inside of the trigram *k'an* (坎卦) at the right side. [123] Inside [the trigram] *k'an* (坎) there is the perfect *yang* (*chen-yang* 真陽). It has the designations *yang*-thunder (yang-lei 陽雷), joy of the heaven (*t'ien-hsi* 天喜), fire-thunder (*huo-lei* 火雷) and Six-and-One (*liu-i* 六一). [124] [The trigram] *k'un* (坤) stands for the [number] Six and the centre. *Ch'ien* (乾) and [the element] metal have the [number] One. [125] [The number] One gives birth to [the element] metal that [in turn] gives birth to [the element] water. Water makes

---

120 It says literally: "my mouth". In the commentaries, we generally could speak in the first person ("I"). However, I prefer the second person ("you") to convey the ideas of a didactic address.

121 I cannot exclude that some other unknown text by Wang Wen-ch'ing (王文卿) originally preceded the commentary.

122 This means that in the sequence of the trigrams the fourth position ("transmission") after "heaven" (*ch'ien* 乾) is the trigram *chen* (震) that therefore is associated with number 4. See *Chou-i*, 9 (ed. Kanbun taikei vol. 16, 1913). Compare J.Blofeld: I Ching, The Chinese Book of Change, pp.216-219.

123 *Yü yu-chih k'an* (於右之坎). Studying the position of trigram *li* (離) we find that the trigram *k'an* (坎) is on the right side, compare J.Blofeld: I Ching, p.217 (King Wen's arrangement).

124 Compare TT 1220: 69.24b, for an interpretation of these terms by the commentator Yü-feng (御風), see below the chapter The Creative Impetus. The astronomical term t'ien-hsi (天喜) may refer to the calendar. I do not think this is the point in our text.

125 Concerning the number see, *Book of Changes*, p.741 (Key for Identifying the Hexagrams); and J.Blofeld: I Ching, The Chinese Book of Change, pp.216, 217.

pools, and pools hold water. It is the water, which makes the rain. This is a most celestial, a most secret and most subtle principle.

When the sound of wind and thunder becomes clearly discernible on the sides of your ears, you expel coughing the breaths nine times. The [number] nine is the old-*yang* (*lao-yang* 老 陽). It is the principle that the zenith of *yang* (*yang-chi* 陽 極) produces *yin* (陰). Now, you realise again nine times the internal practice as described above, and then you feel that the water of the two kidneys has actually ascended, and the water is the rain. When the urine becomes pressing, it is important that you must not release the water of the kidneys when the rain has not [yet] fallen. **(21a)** Just wait straightaway for the wind and the rain to become very strong, and then you [can] get up and slowly release the urine, but then it is that the big rainfall has already come.[126] Only this internal practice and conduct (*hsing-ch'ih* 行 持) never result in even one failure. Can you really be careless about this matter?"

The text clearly presents an equation between the natural phenomenon "rain" in our visible world and internal processes within the body of the priest who guides and organises what we may call the internal rain. We can understand this internal practice as application of the general and familiar concept of using the human body as a microcosm that parallels the macrocosm of the visible world. However, they are knit together in an osmotic relationship that the formal symbols of the eight trigrams feature. The human body being a cosmic mirror functions the same time as an irresistible stimulus for nature. The systematic and rather intellectual approach that we find in these texts characterizes many sources of Thunder Magic that date from the Sung period.

Killing and Crusading (sha-fa 煞 伐), the Martial Side of Thunder Magic

Praying for rain, stopping inundations and other similar good deeds brought the Taoists fame and honours. They believed that natural disasters and deprivations were due to the unruly behaviour of baleful spirit forces that may inhabit official and unofficial shrines. The specialist of Thunder Magic is supposed to know how to handle the problem. Wang Wen-ch'ing (王 文 卿) offers his insights

---

126   In other words, it rains as much as it rains within the body of the practitioner.

concerning the ways and means to tackle the difficult task in the chapter Killing and Crusading (*sha-fa* 煞 伐). We read the following expositions: [127]

(21a) Please let me, [Yüan Wu-chieh (袁 無 介)], raise some questions about the essential requirements for killing and crusading.

[Wang Wen-ch´ing (王 文 卿) gives the following answer:] It is all about the breath of rage (*nu-ch´i* 怒 氣) of the own original spiritual force (*tzu-chi yüan-shen* 自 己 元 神). Any summoning of [spirit] generals and any writing of amulets requires the breath of rage that goes with an urgent and enraged [demeanour]. When you write an amulet, you do it urgently and speedily. Having done the writing you take away the brush [moving the hand towards the direction of] the *sun*-window (巽 戶).[128] At the same time the [spirit] generals and marshals that you summoned to come, all show an enraged apparition. When you dispatch the [spirit] generals, it is necessary that you are tranquil and firm and settle down sitting cross-legged. You concentrate in meditation on the one spark of golden radiance within the spleen that separates to be two sparks. One spark moves on revolving to enter the space between the two kidneys where it changes to become an agglomeration of black breath (*hei-ch´i* 黑 氣). The agglomeration rapidly rushes straight upwards to exit from both eyes. On *yang*-days (*yang-jih* 陽 日) [the agglomeration] exits from the left eye, and on *yin*-days (*yin-jih* 陰 日) from the right eye and soars up (*fei-shang* 飛 上) then towards the *sun*-window (巽 戶). After that, you concentrate in meditation on the [second] spark of golden radiance that revolving enters **(21b)** the heart where it changes to become red breath (*hung-ch´i* 紅 氣). The red breath exits from the base of the tongue and soars [up] towards the *sun*-window (巽 戶).

It is the black [breath], which makes the black clouds and the black mist. It is the red [breath], which makes the fierce fire and the glowing-red shining. You only see the black clouds and the blazing fire that inundate heaven and densely fill the cosmos (*liu-ho* 六 合). Black clouds and glowing-red fire are everywhere around and there is just nothing that they would not burn or cover. Divinities wail and demons cry, and there is no location at all where they could escape to and hide away. However, the black breath encloses the red breath and forms countless agglomerations beyond any limits (*pai-ch´ien-wan t´uan* 百 千 萬 團). At this point of time, a great heat is inside the human body and sweat perspires. You definitely must not move a fan or open your shirt but keep on until the

---

127 TT 1220: 69.21a-23b
128 This always indicates the direction of Southeast.

heat becomes very scorching. You use in a very gentle way lamp-wicks (*teng-hsin* 燈心[草]) [129] to push them into your nostrils (*ch'u-pi* 搐鼻). On *yang*-days (*yang-jih* 陽日) you push a lamp-wick into the left nostril, and on *yin*-days (*yin-jih* 陰日) you push it into the right nostril. You do sneeze once and loudly, and within a moment, just one sound of a thunderclap comes from the window of *sun* (*sun-hu* 巽戶). When you have a response like this, then you can dissolve the [Thunder] altar. Certainly, there will never be any failure at all.

The commentary of Yü-feng (御風) says: Killing and crusading necessarily require that you enrage your outer appearance, your original spiritual force, the fire of your heart **(22a)** and your internal disposition in order to make them utmost hard and enduring just like the movements of the heavenly bodies, which are strong by themselves and never take a rest.

Generally, when you write amulets you must enrage (*fen-nu* 奮怒) the essences and spiritual forces in a determined way, so that they leap up boldly. You must be urgent and quick when you are writing amulets. You use both hands to form the [*mudrâs*] Thunder Office (*lei-chü* 雷局) and thrust them onto [the amulet]. Quickly take off the writing brush [moving your hand towards the direction of] the *sun*-window (*sun-hu* 巽戶). The generals and marshals that you summoned to come all show an enraged appearance. Having dispatched the generals you conveniently sit down cross-legged and practice silent meditation. Your nostrils inhale the pure breaths of the Anterior Heaven (*hsien-t'ien ch'ing-ch'i* 先天清氣). [The pure breaths] force directly their way to the yellow court (*huang-t'ing* 黃庭). The nine orifices above and below are all closed and do not allow any breaths to be absorbed (*jung* 容). Being in a state of tranquillity you let the one spark of golden radiance that is within the spleen ascend and separate to be two [sparks of radiance]. [130] One spark moves on to enter the two kidneys behind the naval, where it becomes an agglomeration of black breath (*hei-ch'i* 黑氣). The agglomeration forces its way upwards from the spine and the double pass to reach the two eyes where it exits. On *yang*-days (*yang-jih* 陽日) [the agglomeration of breath] exits from the left eye, and on *yin*-days (*yin-jih* 陰日) from the right eye in order to soar upwards to the window of *sun* (*sun-hu* 巽戶). You concentrate your mind on the one sound of thunderclap that [then] greatly shakes. After this and for a second time the nostrils inhale the pure breath that forces its way to enter the mysterious female (*hsüan-p'in* 玄牝). [131] Do not allow

---

129   See Herbert A. Giles: A Chinese-English Dictionary, nr. 10864.
130   Literally: „into two parts".
131   The statement concerning the nine orifices is omitted, compare above.

any breath to be absorbed (*jung* 容) and keep the colon (*ku-tao* 穀 道) **(22b)** completely contracted (*i-so* 一 縮). You let the [pure] breath directly ascend to enter revolving the orifices of the heart, where the [pure] breath becomes the red breath that [finally] exits from the base of the tongue and soars upwards to the *sun*-window (*sun-hub* 巽 戶).[132]

[The breath] that is black becomes the black clouds and the black mist. [The breath] that is red becomes the red shining and the blazing fire. You only see the black clouds and the shining of fire that collides with heaven and fills overflowing the cosmos. It burns all around the world-of-the-law (*fa-chieh* 法 界) to clean away any poisonous atmosphere. The wicked spirits find their tracks cut off. Fiery clouds and killing breaths cover and stay at all the locations where wicked forces reside. The ritual officer (*fa-kuan* 法 官) in his meditation directs [the fiery clouds and killing breaths] to all these places. Therefore, it is that the divinities wail and the demons cry in those wicked places. There is just no way for them to escape and hide away. The black breaths enclose the red breaths and form agglomerations. One [agglomeration] transforms to become two [agglomerations] and two of them become three [agglomerations] and the three [agglomerations] become countless agglomerations without any limits.

At this time, a great heat is within your body and sweat perspires. You must not move a fan or unbutton your shirt but keep on until the heat becomes scorching. Gently use lamp-wicks (*teng-hsin* 燈 心 [草]) to push them into the nostrils. On *yang*-days (*yang-jih* 陽 日) you use this method for the left nostril, and on *yin*-days (*yin-jih* 陰 日) you use this method for the right nostril. You sneeze with one [strong] sound that [corresponds with] one shaking thunder. The wicked forces find their tracks cut off, which is such a subtle matter that I just cannot say it.

[The methods of] Killing and Crusading, Praying **(23a)** for Rain and [the methods of the] Creative Impetus (*tsao-hua* 造 化) are identical.[133] You first concentrate your meditation on the red breath in your heart. The red breath forces its way downwards to the mysterious female (*hsüan-p'in* 玄 牝). The nostrils again inhale pure breath that you let expand and collide with the spiritual force of the spleen (*p'i-shen* 脾 神). The one spark of golden radiance that is just below the spiritual force of the spleen makes its way directly to the two kidneys. You first contract [and keep closed] the colon, lift up the black breath of the

---

132   The *sun*-[window] (巽) stands for the nostrils.
133   This statement refers to the chapters nrs. 6 and 8 of this text (TT 1220: 69).

right kidney and let it very slowly move directly upwards to reach the Mud Pill [Palace] (*ni-wan* 泥 丸[宮]). The black breath thus fills overflowing the cosmos (*liu-he* 六 合). The ascent and the absorption of the breaths (*hsi* 息) happen thirty-six times. Now, you let the red breath of the left kidney ascend to move upwards and reach the Mud Pill [Palace] (*ni-wan* 泥 丸 [宮]). When the black breath [there] faces the red breath, they enclose each other and soar up towards the *sun*-direction (*sun-fang* 巽 方), and all the heaven is filled with black breaths and red breaths. A second time you let the black breath of the right kidney ascend, and [again] you breathe in [breath] (*hsi* 息) thirty-six [times] that forces its way up to the window of the brain (*nao-hu* 腦 戶).

Above in heaven and below on earth and absolutely everywhere, there are these black and red killing breaths. They cover and pin down the wicked divinities (*hsieh-shen* 邪 神). Divinities wail and demons cry. They do not have any place where they could escape to with their [outer] apparitions. You grasp the killing breath of the heavenly *kang* [-star] (*t'ien-kang sha-ch'i* 天 罡 煞 氣) that burning blazingly shoots forth from the black breaths. Moreover, your nostrils inhale pure breaths that rush to the Heavenly *Kang* [-star] of the Fire of the Heart (*hsin-huo t'ien-kang* 心 火 天 罡).[134] They rush directly to the T'ai-i Cave (*T'ai-i hsüeh* 太 乙 穴).[135] You use the nostrils **(23b)** to lead on these breaths. When you suddenly release a very heavy load of [these] killing breaths there will definitely be a [fitting] response. The mysteries are reserved for the oral transmission from the teacher master.

The last paragraph of the commentary by Yü-feng (御) contains quite a few terms that are very ambivalent, addressing astral names or elements, which obviously are also human properties. The associative method is a common feature in Taoist sources, and Thunder Magic is no exception at all. Sometimes the actual identification of specific terms must follow the explanations that the respective authors found necessary to give. Some Taoist authors reveal their own understanding of the terms they use. In other cases, we are not so fortunate. Any search for explanations in other and, very likely, later texts is a risky endeavour. It is for this reason, that I did not comment here on statements that explicitly rely on oral instructions to become intelligible. Yet, our texts tell us enough to figure out the basic conditions of Thunder Magic.

---

134 Concerning this phrase, see the following translation of a section of *Hsien-t'ien i-ch'i huo-lei Chang shih-che ch'i-tao ta-fa* (先 天 一 氣 火 雷 張 使 者 祈 禱 大 法). See above pp.8a-8b, and 6a (concerning "the mysterious female", albeit with a quite different application).

135 See below for a description of this term, following TT 1220: 82.18b-19a.

The collection *A Corpus of Taoist Ritual* contains quite a number of complete sets of ritual instructions that address individual divine patrons (*chu-fa* 主 法). They usually include the respective pantheon of spirit-generals in a paragraph with the title General Class (*chiang-pan* 將 班). Listings of the subordinate deities follow.[136] We also find varied instructions concerning ritual matters. For example, we find paragraphs that feature the writing of amulets or the construction of altars that involve the Thunder Divinities. The Great Method for Prayers to the Emissary Chang of the Fire Thunders of the One Breath of the Anterior Heaven (*Hsien-t'ien i-ch'i huo-lei Chang shih-che ch'i-tao ta-fa* 先 天 一 氣 火 雷 張 使 者 祈 禱 大 法) [137] is such a complete set of instructions concerning Thunder rituals under the patronage of Emissary Chang. Here we find a rather instructive tract concerning the Heavenly T'ien-kang [Stars] (*t'ien-kang* 天 罡): [138]

(18a) As to *kang* (罡)[139], the four Correct Ones (*ssu-cheng* 四 正) constitute the bowl of the dipper (*kang* 罡) that occupies (*ch'ü* 取) the correct centre of the four directions, and this is my heart (*wu-hsin* 吾 心). The North has the *tou* [-constellation] (斗) of *yin* and *yang* (陰 陽).[140] Heaven uses the *tou* (斗) to manage the ten thousand transformations. Humans use their heart to dominate and display their affairs. It is for this reason that the heart has seven orifices to correspond (*ying* 應) to the Northern Dipper (北 斗). I demonstrate why it is that [the star] *p'o-chün* (破 軍) [141] is a heavenly *kang* [-star] (天 罡). In my heart, an orifice inside holds muddy blood. When I am enraged my face turns red, **(18b)** and this is so because this blood gets up to my face. The common phrase

---

136 For example, TT 1220: 82.1a-1b.
137 TT 1220: 82.1a-29b
138 TT 1220:82.18a-18b
139 H.A.Giles: A Chinese-English Dictionary, nr.5906 gives the translation "The name of certain stars", pointing to t'ien-kang (天 罡). This means "four stars that form the bowl of the 斗 [nr.11427] constellation,...".
140 *tou* (斗) of course means "peck" or simply a dry measure, and indeed, we could understand the word in this sense: "The North has the measure of *yin* and *yang*". See H.A.Giles: A Chinese-English Dictionary, nr.11427. Generally see TT 629 *T'ai-shang pei-tou erh-shih-pa chang ching*, and see K.Schipper, in: *Companion* p.954, saying that is is a "comparatively late and popular text" that ows much to TT 622 *T'ai-shang hüan-ling pei-tou pen-ming yen-sheng chen-ching* (possibly Sung period).
141 This is a name for the seventh star of the Northern (Big) Dipper. See TT 629 *T'ai-shang pei-tou erh- shih-pa chang ching* 18b-19b.

is: the fire of the heart rises. [The word] *kang* (罡) means *kang* (剛) hard [142] and *ch'iang* (強) strong. [143] Rage then is *kang* (罡) [which means] hard. [144]

As to the heavenly *tou* [-constellation] (斗), the sphere of *yin* (陰) ranges from the star *shu-hsing* (樞 星) to the star *k'uei-hsing* (魁 星). The sphere of *yang* (陽) ranges from the star *shu-hsing* (樞 星) to the star *p'iao-hsing* (飄 星). When the heart is tranquil, it is yin (陰). When the hart is excited, it is *yang* (陽). My heart then is inside [my body the star] *lien-chen* (廉 貞), and on the outside my heart is [the star] *p'o-chün* (破 軍). Now, in the state of tranquillity, I contain my heart inside, and any matters do not bud. [My heart] is in the highest state of being honest (*lien* 廉) and pure (*ch'ing* 清). Outside [of my body] my heart then is the [star] *p'o-chün* (破 軍). When a rage develops, my heart is beating fast and is getting bold as if it were rushing in a hurry. Even if there were legions of armed forces (*chün* 軍), [my heart] could crush (*p'o* 破) them... [145]

This paragraph shows nicely how the association of words and characters constructs the persuasive combination of external, astronomical and internal bodily realities. The ambiguity of many technical phrases that we encounter can also be shown for the term *t'ai-i hsüeh* (太 乙 穴). We already had some statement by Yü-feng (御 風) that used the term. Yü-feng (御 風) commented on Wang Wen-ch'ing's (王 文 卿) exposition Killing and Crusading (*sha-fa* 煞 伐).[146] It is well known that the name T'ai-i (太 乙) refers to a key astral position and stellar deity of antique provenance.[147] The chapter Great Method for Prayers of the Emissary Chang of the Fire Thunders of the One Breath of the Anterior Heaven (*Hsien-t'ien i-ch'i huo-lei Chang shih-che ch'i-tao ta-fa* 先 天 一 氣 火 雷 張 使 者 祈 禱 大 法) in *A Corpus of Taoist Ritual* has a telling tract concerning the term and name of T'ai-i (太 乙), which goes with the taboo name Hun-ming (混 明): [148]

---

142 or „unyielding"
143 or „forceful"
144 See H.A.Giles: A Chinese-English Dictionary, nrs.5895 and 5906 as interchangeable characters.
145 The name of the star means „to crush enemy forces". See TT 629: 17b-19b.
146 See above, TT 1220: 69.21a-23b
147 See F.C.Reiter "The Discourse on the Thunders", p.224. Concerning this theme, see W.Eichhorn: *Die Religionen Chinas*, pp.114-115, 124-127. Stuttgart 1973. Also see, H. Maspero: Le taoisme et les religions chinoises, p.535 (Les procédés de nourrir le *principe vital*). Paris, rpr. 1971. I.Robinet: *Méditation taoiste*, pp.205-210. Paris 1979. Compare TT 1220: 76.22a.
148 TT 1220: 82.18b-19a. The name may mean either "United Luminescence" or "Brightness in Chaos".

(18b) Now, T'ai-i (太乙) is the cave (hsüeh 穴) where yin (陰) is all united. It is the breath of perfect yin (chih-yin 至陰). The cave is exactly in front of the colon (ku-tao 穀道) and behind the urinary organs where it sinks right into the centre [between them]. The [perfect] breath bends the head and stretches the belly. When you contract [and shut up] the two caves (hsüeh 穴),[149] **(19a)** the breath ascends spontaneously. This is the subtle function [that is called] to roll up the water and seize the dragon (chüan-shuei ch'in-lung 捲水擒龍)".

These cryptic lines demonstrate in a way that physical dispositions can be projected to identify implicitly astral positions. Thunder Magic rituals reflect outwardly, what the performing priest manages to evolve in his body, in his mind and in his hands. I remind of the plain connection between the urine that is withheld and the rain that has not yet come. The performing specialist of Thunder rituals is much more than an intermediary. He identifies himself in a practical way with the cosmos and creation, and thus becomes its transcendent body. A small portion of a commentary by Yü-feng (御風) can again illustrate this disposition of Thunder Magic:[150]

(25a) Heaven has the sun, the moon and the stars that can shine upon all the surroundings. Man has the eyes, the ears and the nostrils and can hear, see and unite the essences, the vital energies and the spiritual forces (ching, ch'i, shen 精氣神) in his human body, in the middle between the two kidneys. The left one is the sun, the right one is the moon, and the central palace (chung-kung 中宮) is the heavenly stars t'ien-kang (天罡) that are the ancestral vital energy (tsu-ch'i 祖氣). The human body has the three rivers (san-chiang 三江), the five lakes, the four seas, the five holy mountains and the four marchlands that become apparent in the facial section [of the body]. My body then is great earth. Mountains and rivers are all completely present. Man is able to arrange the breathing and harmonize the vital energies to unite them in his body. He is able to respond to and unite with the mysteries of heaven and earth. Now, I use all my sincere will (ch'eng-i 誠意) to infuse it to my thoughts and so proceed to turn the handle of the dipper (t'a-fan tou-ping 踏翻斗柄).[151] When heaven and earth become unexpectedly dark, this definitely is a response [to this action].

---

149 This obviously refers to the colon and the urinary organs.
150 TT 1220: 69. 25a
151 For this phrase, see above chapter 2 (chao-ho 召合).

The Creative Impetus (tsao-hua 造 化)

The superb position and operational basis of the Thunder specialist becomes again clear when we read still another telling exposition by Wang Wen-ch'ing (王 文 卿). The chapter Creative Impetus (*tsao-hua* 造 化) fits perfectly well to the preceding expositions in this book that speaks a lot about the relationship between human nature and Thunder Magic. I present the first part of the text by the priest Wang Wen-ch'ing (王 文 卿) that actually is the basis for the commentary by Yü-feng (御 風). I used the commentary to conclude the preceding paragraph:[152]

(23a) Yüan Wu-chieh (袁 無 介) says: Heaven and earth create and bring about thunder, lightning, wind and rain. What causes could make them occur?

[Wang Wen-ch'ing (王 文 卿) gives the following answer:] Man is endowed with the breath of heaven and earth, and he lives on it. Heaven and earth are correct and selfless. Man, however, can snatch the creative impetus of heaven and earth. Now, the breaths of the three potentials (*san-ts'ai* 三 才), heaven, earth and man are omnipresent and [the same time] they are individual breaths.[153] All causes are together with heaven and earth mutually inside and outside. As to wind and rain, thunder and lightning, why should it be difficult to make them occur?

Now, heaven has the sun, the moon and the stars. Their radiance is bright enough to shine on the world all around. Man has his eyes, ears and nostrils, and so he can hear, see and know all creatures between heaven and earth. Earth has the three *chiang* [-rivers] (江), the five lakes, the four seas, the five holy mountains and the four marshlands, and there are four limbs for all creatures. The [human] body accordingly is **(24a)** great earth, and the mountains and rivers are all completely present. I actually can put an end to the causes (*hsi-yüan* 息緣) and harmonize the breaths (*t'iao-ch'i* 調 氣) in order to take the adequate responses from within my body and unite [myself] with the secrets of heaven and earth. Therefore, it is that using my own perfect will (*chen-i* 真 意) I focus my thoughts on the ongoing [ritual] procedures (*suo hsing chih shih* 所 行 之 事).

---

152   TT 1220: 69.23b-27a. In this section I skip the commentaries.
153   I do not think that *p'ing-hsi* (屏 息) means "abstain from uttering any sound". See H.A. Giles: A Chinese-English Dictionary, nr.9319. Here, the term indicates the individual breaths of each of the three potentials. *P'ing* (屏) basically means "to set aside".

The perfect breaths of heaven and earth thus follow my will and action. I fix my [internal] vision and seize the response, which never fails in even one single case.

Wang Wen-ch'ing (王文卿) continues his instructions on **p. 25a:**
"Now, what is born out of me, this is my will. Good and bad, they all emerge from my will. The will is the spiritual agent of the spleen (*p'i-shen* 脾神). The spleen belongs to [the element] earth. The number of its coming to life (*sheng-shu* 生數) is the Five, and the number of its completion (*ch'eng-shu* 成數) is the Ten. The outer shape of the spleen resembles the crescent moon (*yen-yüeh* 偃月) **(25b)**. When there are no thoughts and calculations the two sharp points [of the spleen] hang downwards. When all the thinking (*i-nien* 一念) suddenly arises, the two [sharp] points consequently rise and point upwards. Any single spark of thought and calculation emerges at the centre of the spleen that in fact is the ancestral breath (*tsu-ch'i* 祖氣) [154] of heaven and earth. As soon as any single thought is born, it floats (*ku* 泪) into the heart. Now, the heart is the lodge of the spiritual agents. When the heart receives this one single spark of the will, which is good or bad, vile or killing, which may have [the quality of] an immortal or of a Buddha, the heart serves them all.

If I am able to know the source and the flow (*yüan-liu* 源流) of the ancestral breath, I must resolutely hold on to this one spark of perfect breath (*chen-ch'i* 真氣). I am (*wo* 我) then [in the position of] ruler (*chu-tsai* 主宰) [155] in heaven and on earth, which pertains to the employment of my heart. If I let my heart be enraged, cruel and bad, I lack the Tao of Heaven (*t'ien-tao* 天道) and shall eternally be a demon of the lowest class (*hsia-kuei* 下鬼). How could I still be able to seize the creative impetus of heaven and earth?" [156]

These texts show clearly that the terminology of internal alchemy (*nei-tan* 內丹) describes the fundamental conviction of the existential identity of man and the cosmos. This notion implies the precondition of a human identity with those divine potentials that carry life. The external thunder and rain appear to be functions and expressions of thunder and rain within the human body, where they first have to be aroused. The cosmos inevitably will mirror and echo such internal processes.

---

154 Or say "ancestral vital energy", which I eventually use as alternative translation for the sake of convenience.
155 I notice that H.A.Giles: A Chinese-English Dictionary, nr.11490 offers the translation "god".
156 „...and act as a ritual specialist?" This is the implied question.

We find that the texts by Wang Wen-ch'ing (王文卿), especially his *Wang Shih-ch'en ch'i-tao pa-tuan chin* (王侍宸祈禱八段錦) show a rather modest application of *nei-tan* (內丹) terminologies.[157] The commentaries on the other hand indulge in *nei-tan* (內丹) rhetoric, which was in tune with the culture of the time. Comparing the School Talks with this text "in eight paragraphs", we hardly can claim that those specialists of Thunder Magic fostered a common and mandatory set of *nei-tan* (內丹) terminologies. The terminology and rhetoric may have depended on the individual affiliation of the respective priest and his education. We know that many materials went through the hands of later compilers and editors, and Pai Yü-ch'an (白玉蟾 13th ct.) certainly was only one of them. It is risky to project later interpretations backwards and claim them for an interpretation of earlier works. I certainly want to repeat that there is no need to do so, although we may not be able to understand every coded indication.

Thunder Magic was a very practical matter. I continue to present a chapter of *A Corpus of Taoist Ritual* that we connect with some certainty with Wang Wen-ch'ing (王文卿). It contains theoretical and practical instructions that show the range of ritual approaches and methods in Thunder Magic. However, there is no connection with the cult and the ritual patronage of any specific Thunder divinity. The texts contain general and didactic instructions that serve us well.

I already said that *A Corpus of Taoist Ritual* contains many ritual programs that are based on the patronage of specific deities. The programs include the lists of the pantheon and instructions about ritual steps, amulets, *mudrâs*, spirit hells and other ritual devices that may be used to control the spirit forces. I call to mind, for example, the Great Method for Prayers to the Emissary Chang of the Fire Thunders of the One Breath of the Anterior Heaven (*Hsien-t'ien i-ch'i huo-lei Chang shih-che ch'i-tao ta-fa* 先天一氣火雷張使者祈禱大法).[158]

The following paragraphs in this book present the comprehensive and informative Divine Texts of the Great Methods of the Five Thunders at the Jade Department in the Heaven of Highest Purity (*Shang-ch'ing yü-fu wu-lei ta-fa yü-shu ling-wen* 上清玉府五雷大法玉樞靈文).[159] This text deals with many practical devices of Thunder Magic that all have transcendent foundations, which is well explained in the detailed preface by Wang Wen-ch'ing (王文卿).

---

157   TT 1220: 69.1a-27b
158   TT 1220: 82.1a-29b, see above.
159   TT 1220: 56.1a-42a

He gives a survey on the divine potentials that condition the pantheon and the various means of the Thunder specialist. He adopts in his presentation the position and authority of the evasive figure of his teacher who allegedly was the Fire Master of Thunder and Thunderclaps. He displays the Arrangement of the Offices of Thunder and Thunderclaps together with an extended Thunder pantheon.[160] Some paragraphs of the text present the Precious Seals of Thunder Might and the methods to use them. Much attention is being paid to the Locations and Departments of the Five Thunders and their different types. We find again, in a way unexpected, the story of the Great Divinity of Law and Order who closely links Thunder Magic with the national legendary history.[161] The application of Thunder breaths, the writing of amulets to cure illnesses, to raise dragons and produce rainfall, such activities are major themes in *chapter 56*. Spells, *mudrâs* and ritual steps all matter greatly in these instructive texts. How to build thunder altars, how to construct hells to put away demonic forces and how to crusade against evil deities in their shrines, such questions and many more get fascinating answers in *chapter* 56 of *A Corpus of Taoist Ritual.* They always remind us of the explanations that we have just seen.

The priest moves within the solid frame of his religious career, receiving registers and oral secret instructions. In the case that he performs bravely his tasks, he can rise in the spirit administration up to divine functions and ranks. This seems to ascertain the basic conviction of the divine nature of the priest, which he can consciously employ in ritual.

Yet, some Thunder rituals are performed with "dishevelled hair and barefooted". There is no word about robes and ritual caps or an especially dignified demeanour. The demeanour of the Taoist Yeh Ch´ien-shao (葉千韶) reportedly was not always very dignified and in many ways may remind us of the modern exorcists whom we have seen in action in Taiwan.[162] However, we have to accept, I believe, any Taoist self-identification. We also should accept what Taoists of any epoch tell us about the range of their theoretical and practical ways and means and the constituents of their religion and culture.

---

160   TT 1220: 56.5a-10a. The pantheon comprises more than one hundred abstract names that I do not list.
161   TT 1220:56.14b-15b. See above "Assembling the Divine Force", in: TT 1220: 124.1b-2a.
162   See above my Introduction. Compare P. Nickerson: "Attacking the Fortress, Prolegomenon to the Study of Ritual Efficacy in Vernacular Daoism", pp.117-179, in: AAS 20.

All our themes concerning the theoretical and practical elements of Thunder Magic reappear in the many other chapters of *A Corpus of Taoist Ritual*, albeit with many variants that are due to different historic and regional backgrounds. Due to the introduction by Wang Wen-ch'ing (王 文 卿) *Chapter 56* shows a rather comprehensive and authoritative documentation of the historical and yet general conditions of Thunder Magic.

Chapter II: The Scope of Thunder Magic

A Corpus of Taoist Ritual (Tao-fa hui-yüan 道法會元) contains quite a number of comprehensive sets of Thunder rituals. They often represent regional traditions and affiliations that we hardly can trace concerning their historic roots. Fortunately, *chapter 56* in *A Corpus of Taoist Ritual* has a preface by Wang Wen-ch'ing (王文卿) that at least suggests that the chapter has his spiritual patronage. I present most of *chapter 56* in translation to complement the preceding presentation in Chapter I. The title of *chapter 56* in *A Corpus of Taoist Ritual* is promising: [1]

Divine Texts of the Great Methods of the Five Thunders at the Jade Department in the Heaven of Highest Purity, (*Shang-ch'ing yü-fu wu-lei ta-fa yü-shu ling-wen* 上清玉府五雷大法玉樞靈文),

Preface by Wang Wen-ch'ing (王文卿):

(1a) The Fire Master of Thunder and Thunderclaps (*lei-t'ing huo-shih* 雷霆火師) says: formerly when Heaven and Earth (*hsüan-huang* 玄黃) were not yet separated, [2] there was a phase of dimness with hardly any light at all. When the clear and the turbid then separated, the existence of the visible signs of the Vital Principle (*hung-meng* 鴻蒙) [3] came first into being.

The Supreme God-Emperor of Prime Origin (*Yüan-shih shang-ti* 元始上帝) saw that there were no principles for symbols and signs, and so he initiated the great saintly beginning with his command to carve and define the opportune and clear cut rules and standards for the form of the characters. The precious books on flowery writing tablets, the dragon texts and phoenix seal characters thus became manifest in radiant light.

At that time, the Perfect King of Jade Purity (*Yü-ch'ing chen-wang* 玉清真王) of the exalted *Shen-hsiao* (神宵) [-Heaven] stayed at the Golden Palace of Congealed Spiritual Might (*Ning-shen chin-ch'üeh* 凝神金闕). He reflected on

---

1 TT 1220: 56.1a-42a
2 *Chou-i* 1 (*k'un* 坤), p. 23, in: *Kanbun taikei* (Tokyo 1913). See *Book of Changes* p. 391.
3 *Chuang-tzu chi-chieh* (莊子集解) 3 (tsai-yu 11 在宥), p. 66, in: *Chu-tzu chi-ch'eng* (諸子集成). (Peking 1986). Compare H.A.Giles: A Chinese-English Dictionary, nr.5269.

the saints and nobles who sympathised with the living beings in the regions beneath where they float drowning in the the sea of bitterness. They drift along in the waves of birth and death. In ten thousand *kalpas,* they experience injustice and error, and the bewitching spirits are able to destroy or harm [them]. The Perfect King felt sorrow and sympathy, and using the proper formal way he questioned whether any diligent attention was being paid [to the situation].

The Supreme God-Emperor of Prime Origin dwelling in the Golden House at the Jade Palace said to the Perfect King:

Those creatures are not yet **(1b)** enlightened and consequently concentrate in themselves the [*karmic*] causes (*she-yin* 攝 因).[4] If the creatures were enlightened about the fact that they concentrate the [*karmic*] causes in themselves, how could they [still] be in the Hell of Earth? As to the Halls of Heaven, would they then not be the befitting places for them?

The Perfect King kowtowed, repeatedly saluted, knelt down for a long time and looking up with respect stated his gratefulness for the heavenly mercy and expressed his vow to hold on to the essentials for perfection. The [God-Emperor of] Prime Origin felt sympathy with him, and was that not good?

Thereupon [the God-Emperor] ordered the Divine Favourite (*ling-fei* 靈 妃), the Jade-girl of Cave Mystery (*Dong-miao yü-nü* 洞 妙 玉 女) that she takes the gilded key that dragons adorn and open the lock with jade inlays of a whistling phoenix at the Temple Subtle Perfection of the Flowers of the West (*Hsi-hua miao-chen tien* 西 華 妙 真 殿). She was to take the chest made of shining jade and precious stones and confer the transmission of the Tao of Primordial Unity (*hun-ho chih tao* 混 合 之 道) of the Purple Script of the Three Luminaries (*san-kuang tzu-wen* 三 光 紫 文). She was to teach [the Perfect King] the *mudrâs* (*chüeh* 訣) of the Five Thunders, which control wicked forces and behead bewitching demons. [The *mudrâs*] have the desigantions Black Dragon, Red Horse, Blazing Fire and Prosperous Circuit. One uses them to dispatch the emissaries from the empyrean, which is a perfectly hidden and most secret essential [matter].

At that time, the Supreme God-Emperor of Prime Origin directed the great lot of complaisant saints of the ten regions to return, and he ordered the demons to

---

4 See W.E.Soothill and L.Hodous: A Dictionary of Chinese Buddhist Terms, p. 205a. (Rpr. Taipei 1972). Also compare Ting Fu-pao (丁 福 保): *Fo-hsüeh ta tz'u-tien* (佛 學 大 辭 典) 984c-985a (*shih-yin* 十 因). Taipei 1974.

ascend and enter the state of formlessness. The Perfect King was sent down to stop the evil world. The Perfect King thereupon took the amulets and seals, the dragon-tracts of the Five Thunders **(2a)** and the secret instructions of the astronomical calculations concerning the Northern Dipper as to transmit them to the Elder of the Five Holy Mountains (*Wu-yüeh chang-jen* 五嶽丈人) [5] saying:

It is the Northern Constellation (*pei-ch'en* 北辰) of Heaven, [6] which all the other stars salute.[7] Below [the constellation] there are the two palaces that are named the Pivot and the Initiatory Force (*shu-chi* 樞機), [8] and they are said to be the two poles of the East and the West. The Palace Pivot revolves and has the might that transforms life (*hua-sheng* 化生).[9] The Palace Initiatory Force is the institution that on behalf of the god-emperors executes the commands and regulates the nine hundred and six [types of] disasters and calamities in the *yang*-[sphere] (陽). [The Palace Initiatory Force] is in charge of all affairs that pertain to happy and cruel [events] among people, to demons and divinities, disastrous and favourable events. It issues and brings to life all the beings with their categories and oscillating appearances. All of this is evidence for orders and commands, and it is for this reason why thunder divinities (*lei-shen* 雷神) exist.

There are five types of thunders, namely the Heavenly Thunder (*t'ien-lei* 天雷), the Divine Thunder (*shen-lei* 神雷), the Dragon Thunder (*long-lei* 龍雷), the Water Thunder (*shui-lei* 水雷) and the Earth-Altar Thunder (*she-ling lei* 社令

---

5 Huang-ti (黃帝) is said to have bestowed this Taoist title on a Taoist who lived on Mount Ch'ing-ch'eng shan (青城山) in Szechwan, see TT 1032 *Yün-chi ch'i-ch'ien* 120: 13a; 122.10b. However, in this text *Wu-yüeh chang-jen* most likely refers to a transcendent divinity who is linked with the very beginning of Taoist revelations. TT 599 *Tung-t'ien fu-ti yueh-tu ming-shan chi* 4b also refers to a much later stage in Taoist history.

6 *Pei-ch'en*, Horus boréal, which is connected with the divinity *T'ien-huang ta-ti*. See G. Schlegel: *Uranographie Chinoise*, p. 815. (Leiden 1875).

7 Or translate: "encircle", see H.A.Giles: A Chinese-English Dictionary, nr. 6575 (*kung* 拱).

8 "Pivot and Initiatory (or "Moving") Force" is an ambivalent term. It can stand for crucial units of the (spirit) administration and military forces. I remind of the term *shu-chi fang* (樞機房) denoting a "Central Control Office" (T'ang), see Hucker, nr.5418. My translation of the term is meant to imply the function of a transcendent military office as indicated. F.C.Reiter: "Discourse on the Thunders", p. 213 (translating TT 1220: 67. 21a, *Lei-shuo* 雷說 by Wang Wen-ch'ing 王文卿), in: JRAS 14/3. Also see the same: "A Preliminary Study of the Taoist Wang Wen-ch'ing (1093-1153)", p.181, in: ZDMG 152. TT 1220: 67.3a explains that the term identifies Thunder and thunderclaps (Chang Shan-yüan 張善淵 : *Lei-t'ing hsüan-lun* 雷霆玄論).

9 In Buddhism, this also is an important term, see W.E.Soothill and L.Hodus: A Dictionary of Chinese Buddhist Terms, p.142a (Anpapâdaka).

雷). Furthermore, there is the Thunder Wall (*lei-ch'eng* 雷 城) that is located amidst the *Brahma*-breaths (*fan-ch'i* 梵 氣) above the white and jade-like milkyway and the Prefecture of the Perfect King in [the Heaven of] Jade Purity. The Thunder Wall is 2300 miles away from the region of the Prefecture of Jade Purity. The Wall is high eighty-one *chang* (丈), [10] and I [11] (*wu* 吾) have the Thunder Wall in control. Due to the resident officials, masters, emissaries, ministers and branch offices for distinct duties there is a leading force for the bestowal of **(2b)** life for all creatures that urges on and moves the sea and the mountains. Thus, the four seasons are pushed on and shifted, *yin* and *yang* are made to rise and descend, and [that leading force] registers the good and punishes the bad. The Five Thunders Headquarter Office is the very centre that right in time issues the orders. The Headquarter has its own Special Offices of the Five Thunders Emissaries that comprehensively assist the Five Thunders. Below, they control the three offices, [named] North Pole (*pei-chi* 北 極), Jade Pivot (*Yü-shu* 玉 樞) and *P'eng-lai* (蓬 萊).

When in all the world wind and rain are untimely and excessive heat causes torching droughts, locusts bring calamity, armoured strive madly wages and famine happens often and severely, all of this is due to requests for [divine] commands. The god-emperors (*ti-chün* 帝 君) [12] issue the orders and the Jade Pivot forwards the commands (*hao-ling* 號 令) [for the execution of the orders]. The pivot itself operates three hundred officials in authorized positions who assist and support the control of the transformations of life (*sheng-hua* 生 化).

All men, who strive for perfection, yearn for *Tao* (道) [but] do not discuss the essential mysteries they will have their outer form recede, deteriorate and decay. [In the case that] they do not use amulet water (*fu-shui* 符 水) their efforts will not reach up to the Three Heavens (*san-t'ien* 三天). [In the case that] they do not rinse [their mouth with] the refined essences (*ching-hua* 精 華), their spiritual forces are not pure and pleased. [In the case that] they do not help people in illness and distress, the success of *Tao* will be hard to achieve.

Concerning those who yearn for the immortals, study *Tao* and wish to be rapidly successful, nothing is better than the beheading of evil forces, the expulsion of

---

10 Concerning the Thunder Wall, see F.C.Reiter: "Discourse on the Thunders", p. 218, and note 90, in: JRAS 14/3.
11 Following the phrasing in TT 1220: 56.3b the [Perfect] King in the [Heaven] of Jade-Purity should be the speaker.
12 TT 1220: 56.4a has *ti-chen* (帝 真) in a similar phrase. Also compare TT 1220: 84.9a "...all comes from orders by the god-emperors".

harm, the application of **(3a)** amulets, the consecration of water (*chou-shui* 祝水), the help for people and profiting the creatures, the ample collection of hidden merits and in subtle ways the support for religious veneration (*hsiang-huo* 香火). When [such persons] come together with creatures, they consider how to save them. They are correct and straight without any wickedness, and they behave themselves so all life long from its very beginning. Now, how should they ponder that the easy ascent [to heaven] could not be achieved?

Just be afraid that you are not able to meditate in a refined way and be afraid that you take resort to a superficial and shallow [demeanour]. [Be afraid that] the amulets and *mudrâs* (*fu-chüeh* 符訣)[13] are incomplete, that merits come late and effects are little. [Be afraid that ritual] engagements do not have any magic force [at all], and in the end your remissness is overwhelming. I am very much embarrassed about the situation, and therefore I take today the secret instructions and the divine texts of the Jade Pivot to hand them down to the world of man.

If the secret instructions and the divine texts of the Jade Pivot are internally applied, they can be used for the cultivation and control of the body. If they are externally applied, they can be used to pacify the people and help the country. [They can be used] to pray for rain, clear the sky, eliminate disasters and do away with unfortunate events. They can be used to drive away dragons and snakes that cause calamities in rivers and lakes and restrain scaly dragons and water monsters in pools and caves where they harm people. [The secret instructions and divine texts] save the living beings (*sheng-ling* 生靈) from the distress of great dangers and release the legions of *hun*-souls (*hun* 魂) that face the darkness of Hades. [The secret instructions and divine texts help] to crusade against temples, expel any wicked forces, eradicate evil spirits and heal sickness. They make the thunders and thunderclaps circulate [starting out] from your palms, and they cause wind and rain to come right before your eyes. [The secret instructions and divine texts help] to adjust the crimes and faults of seven generations of ancestors in the nine dark [regions]. They abolish **(3b)** the blood sacrifices of thousands of lives in hundreds of *kalpas*.

Keep close to the divine texts (*ling-wen* 靈文) and secret manuals and so have the cause to be in charge of the creatures and profit the living beings. The subtle instructions and divine recipes (*shen-fang* 神方) save all around from deadly disasters, from confusion and illness. Those who have received [these

---

13 Or translate: the "instructions for [the composition of] amulets".

documents] must keep them safely. Those who come across them have the destination to do so. They have the [required] restraint and practice self-cultivation (*hsing-ch'ih* 行持). [14] [Such persons] can wait to ascend soaring [to heaven]. Being diligent at the beginning and remiss in the end, [such persons] drown eternally in the see of misery. Perfectly investigate [the magic texts and secret manuals] and set them to use. Echo and response will immediately occur.

T'ai-su ta-fu ning-shen tien shih-ch'en Wang Wen-ch'ing (太素大夫凝神殿侍宸王文卿) The Attendant Wang Wen-ch'ing, the Gentleman *T'ai-su* of the Temple *Ning-shen*, wrote this preface."

(3b) The Arrangement of the Offices of Thunder and Thunderclaps (*lei-t'ing fen-ssu* 雷霆分司) [15]

The Fire Master of Thunder and Thunderclaps (*Lei-t'ing huo-shih* 雷霆火師) says: when the Five Thunders is well understood, you should know all the departments of the thunder prefecture (*lei-fu* 雷府). When they all are known there will be immediate responses in the case that requests were made. Furthermore, the Thunder Wall is located amidst the *Brahma*-breaths (*fan-ch'i* 梵氣) above the white and jade-like milky-way and the Prefecture of the Perfect King in [the Heaven of] Jade Purity. The Thunder Wall is 2300 miles away from the sphere of the prefecture. The Wall is high eighty-one *chang* (丈). It is the location, [16] which the King of the Heaven of Jade Purity has in control. His officials, masters, emissaries, ministers and branch offices for distinct duties rule over the bestowal of life for **(4a)** all creatures, urge on and move the sea and the mountains, push on and shift the four seasons, let *yin* and *yang* rise and descend and register the good and punish the bad. In the very centre, there are the rulers of the Five Thunders who right in time issue the orders to support the operations of the Perfect King. [17] Furthermore, the emissaries of the Five Thunders are specialised officials at the Thunder Wall who comprehensively assist the Five Thunders and relate the orders with the individual offices.

---

14 Or: "internal cultivation". See above the chapter A First Approach to Thunder Magic.
15 This section very much reads like a later summary or even repetition of the preface by Wang Wen-ch'ing (王文卿).
16 Or translate: "office" (*suo* 所).
17 See above the translation of p. 2a-2b. This text closely parallels the introduction by Wang Wen-ch'ing (王文卿).

Whenever wind and rain do not accord with the season and torching heat becomes a burden, armoured strife madly wages and famine happens often and severely, all of this is due to requests for [divine] orders by the god-emperors and the perfected ones (*ti-chen* 帝 真).[18] There is just nothing, which does not come from the Jade Pivot, which in a great way covers the detachments of the regions (*fen-yeh* 分 野). At the same time, it leads on the generals and emissaries of the Three Monitoring Offices (*san-ssu* 三 司) to judge over and regulate the merits and failures of demons and divinities in the three realms (san-chieh 三 界), and thus take care of the wellbeing of the black haired people.

Furthermore, in the *tou*-constellation (斗) there are divine prefectures (*shen-fu* 神 府) and the [Palace] Pivot (*shu* 樞). The Pivot has administrators (*hsiang* 相), and therefore there is a Court of the Jade Pivot (*yü-shu yüan* 玉 樞 院) that is also called Court of the Pivot Constellation (*tou-shu yüan* 斗 樞 院). Officials in authorized positions have distinct duties. They are about two hundred officials who support the administration of the Perfect King and guide the duties of thunder and thunderclaps. Concerning the disaster of floods and drought, armoured strife and famine, they all come from corrective actions and thus [are bound to] happen.

(4b) The Headquarter Office of Thunder and Thunderclaps (*lei-t'ing tu-ssu* 雷 霆 都 司) is the Special Control Office (*chuan-ssu* 專 司) of the God-Emperor of the North (*Pei-ti* 北 帝) that arranges the ranks of the officials, distributes the individual duties and assists the governance of the Jade Pivot.[19] Whenever in the world floods cause inundations [or] drought-demons [operate], in all cases one asks the Court of the Jade Pivot (*yü-shu yüan* 玉 樞 院) that the respective reports [about the disasters] would be transferred to the higher echelons and that action would be taken. As to the battle-axes and halberds of thunder and thunderclaps, as to applause, rewards and punishments, they all have their regulations and are not in a state of confusion. There are officials who are in charge of all of them.

The Heart of Heaven (*t'ien-hsin* 天 心) has the thunders but they are not only there. Furthermore, there is the *P'eng-lai* Office (*P'eng-lai ssu* 蓬 萊 司) that is controlled by the Capital Commissioner of the Waters. His generals and

---

18 See above p. 2b for a similar phrase, which uses the term *ti-chün* (帝 君).
19 The text writes "Jade Initiatory Force" or perhaps "Jade *Chi*-[Military Office]" (*yü-chi* 玉 機), which most certainly is a mistake. This phrase does not occur anywhere else. I suppose that we should read *yü-shu chih cheng* (玉 樞 之 政).

Chapter II: The Scope of Thunder Magic

emissaries are specialized to administer the duties concerning the water. They distribute the clouds, scatter the breaths and equally [take care of matters relating to] the [Ch´ang-]chiang, the sea, the [Huang]-ho, the marshes, the springs and fountains. When excessive heat occurs in the world, you must report to the Court of the Jade Pivot. You ask for and memorialise the request that heavy rain and soaking moisture may be sent down to save the living people.

The Fire Master says: As to the [before mentioned] four offices, only the Five Thunders Court (*wu-lei yüan* 五 雷 院) have the sole, privileged authority (*chuan-ch´üan* 專 權). Although the [four offices] all have their [specific] regulations (*chih* 制), each office has fierce generals, emissaries and troops that comprehensively assist the four courts (*ssu-yüan* 四 院). Their might and divine force is vast and grand. How could they be the same as the martial forces (*ping* 兵) of other departments?

The people who study perfection and receive the ritual methods (*feng-fa* 奉 法) all request [the service of thunder] troops due to these orderly divisions. **(5a)** On the transmission of the [ritual] norms (*ch´uan-k´o* 傳 科) [20] one should also get the proper knowledge concerning the Divine Ranks of Thunder and Thunderclaps (*lei-t´ing shen-wei* 雷 霆 神 位).

An extended paragraph follows **(p. 5a-10a)**: The Divine Ranks of Thunder and Thunderclaps (*lei-t´ing shen-wei* 雷 霆 神 位). The list displays the variety of administrative and military ranks that more than one hundred thunder divinities hold. Almost all of these names describe deified functions of the thunders and go without any indication of personal names. [21]

On page **(10a)**, we read the title: Precious Seals of Thunder Might (*lei pao-yin wen* 雷 寶 印 文): [22]

The Fire Master says: The Thunder Department has six amulet-seals (*pao-yin* 寶 印) [23] that were transmitted for the first time by the Superior God-Emperor of

---

20 This refers to the initiation as a priest and thunder specialist.
21 F.C.Reiter: "The Name of the Nameless and Thunder Magic", pp. 97-116. In the following presentation, I locate some of the divine names in the pantheon that does not list all the names we encounter, and eventually there are some variants.
22 *Yin-wen* (印 文) "Seal" means the material object and also the pattern or script that is incised on it. For this theme see, for example, the following chapter 57 in TT 1220. Seals are a rather prominent matter in sources on Thunder Magic.
23 The numeral for seal can be the word *k´ou* (口). I do not exclude that here *fu* (amulet or

Prime Origin (*Yüan-shih shang-ti* 元始上帝) to come secondly to the Five Gentlemen and Elders of the Five Holy Mountains (*wu-yüeh wu-chün chang-jen* 五嶽五君丈人). These seals were all made of jade and eventually were called Jade Pivots (*yü-shu* 玉樞). Later on, Huang-ti 黃帝, Lei-kung 雷公, Feng-hou 風后 and others in the Taoist tradition (*ssu-fa* 嗣法) carried these seals on their belts and ascended with them up to heaven. Formerly, Lord Hsü Sun (許遜) was the first to receive these five symbols [24] and together with Wu Meng (吳猛) assisted to behead a poisonous dragon.

When the poisonous dragon did not yield, [Hsü Sun] (許遜) used a [thunder] seal to shine on the dragon and then the blood flowed from the two eyes of the dragon that could no longer **(10b)** turn and move. Then it was possible to stab and kill the dragon. The blood and filth of the corps splashed on the seal causing subtle cracks. After [the success] Hsü Sun (許遜) again sacrificed and so the [original] pattern (*wen* 文) [on the seal] was restored. In the centre of the seal, there is the Secretarial Receptionist [25] of the Centre of the *Tou*-Constellation (*tou-chung t'ung-shih she-jen* 斗中通事舍人), the Young Lad of Purple Radiance (*Tzu-kuang t'ung-tzu* 紫光童子). [26] In daily life [27] [the seal] must not be used in any faulty way.

The six seals are all made of the kernel wood from a *yang*-peach tree (*yang-t'ao* 陽桃). You use *ping* and *chia* days (丙甲) to carve the seal. In case that peach trees are not available it is also fine to take jujube wood [at the time when] thunders cause alarm. [28]

As to the six seals, there is the Internal Seal of the Emissaries of the Five Thunders, which none must use who does not hold the rank of Heavenly Official of the First Rank of Prime Origin (*shang-yüan i-p'in t'ien-kuan* 上元一

---

contract, 符) is used in this sense. Thunder seals have certainly the meaning of sealing contracts.
24 The word *hao* (號) points to the specific designation or name of the individual seals. See H.A.Giles: A Chinese-English Dictionary, nr.3884.
25 See *Hucker*, p. 556, nr. 7507.
26 See TT 1220: 56.9a and p.9b where we find the names of some other attendants („lads") that appear somewhere else in this text. The Taoists Hsü Sun (許遜) and Wu Meng (吳猛) are not listed in the pantheon.
27 *ch'u-ju* (出入) could also conclude the preceding sentence. The translation would then read as follows: "… (*Tzu-kuang t'ung-tzu* 紫光童子), who comes and goes". This would mean that the divinity commutes to and from the seal.
28 Unfortunately, the respective character is barely readable but I suppose that it is nr.2148 in H.A.Giles: A Chinese-English Dicitionary. Nr.12789 ("violent") may also be an option.

品 天 官). [29] The following rank has the Seal of the Mercury Heaven of Purple Radiance. Those who receive the seal carry it on their belt. They rely on it for years to use it when there are the occasions to summon thunder divinities, move mountains, exhaust the sea, expel dragons and cause rainfall.

Now, there are hidden forms [of spirit forces] that [either] soar in the void [or] walk on earth. They are fierce and wicked demons and deities, and you just use the seal in order to attach it to the spot where they dwell. They shall spontaneously perish. When you should raise dragons and cause rainfall, you use the seal to seal the amulets and memoranda (*fu tieh* 符 牒). However, you must not use the seal in any faulty way. Divine, indeed, is the seal.

The seal of the subsequent rank is the Seal of the Fire Script of Thunder Radiance. **(11a)** Those who happen to receive the seal can use it in order to employ and dispatch the Five Thunders, punish and behead wicked and malignant spirits, raise clouds, bring about rainfall and subdue water monsters (*shui-kuai* 水 怪). If men who cultivate perfection and concoct the elixirs receive the seal, they can use it to secure the site of an altar (*t'an-ch'ang* 壇 場). Demons and monsters see its red shining over a distance of ten thousand *chang*, all the fire and smoke that range across heaven. The demons and monsters cannot show up [at the site] as intruders and thieves. When you carry [the seal] on your belt and happen to find yourself amidst [fighting] troops, blank weapons cannot harm your body. Concerning amulets and memoranda, you also use the seal to seal them.

Formerly, Lord Mao (Mao-chün 茅 君) carried the seal on his belt. His rank was promoted to be Superior Chamberlain [30] and Arbiter of Fate at the Holy Mountain of the East (*Tung-yüeh shang-ch'ing ssu-ming* 東 嶽 上 卿 司 命). Having ascended to heaven, he later bequeathed the pattern of the seal (*yin-wen* 印 文) to our world.

Then there is the Seal of the Cavern-Magic of Jade-Dawn (*yü-ch'en tung-ling* 玉 晨 洞 靈) that is also called Seal *Tou-k'uei* (斗 魁 印). Formerly Lord Mao additionally carried the seal on his belt. Anyone who happens to get this seal can use it to shine on and behead scaly sea serpents and wicked dragons. He can use it to summon thunder and cause rainfall. The immediate responses happen right in time, and [the seal] never fails. In case that you should pray for a clear sky, for

---

29 This is the divine title of the priest, who holds such seals.
30 See *Hucker*, p. 407, nr.4987.

snow to fall, to transmit salvation (*ch'uan-tu* 傳 度)[31] and dispatch [divine forces], for all these activities, you use the seal to seal [the appropriate documents]. Perhaps you issue formal documents, memorialise petitions as well as all the other types of official memoranda, amulets and reports, they shall ascend and arrive in time [in the heavens]. Generally, **(11b)** when you have these seals on your belt and your way takes you to provinces and districts, where religious rites[32] for the god of the city wall and moat take place (*Ch'eng-huang* 城 隍) you are always respectfully received.

(11b) Method of Consecrating Seals (chi-yin fa 祭 印 法)

For any consecration of thunder seals, you take six *ping* (丙) days when thunders had occurred. You arrange at night and beneath the pivot of the dipper an altar table that faces the *kang*-stars (罡). You put some fruits of the season on the table, but you must not use pomegranates, water chestnuts, the roots of lotus, *ko*-beans (葛), sweet cane, black cucumber (*t'u-kua* 土 瓜) and the like. You properly take twelve portions of tee and wine, burn incense and look far out to the pivot of the dipper. You offer repeatedly your veneration to address above [in the heavens] the Commissioner at the Dipper and Secretarial Receptionist (*tou-hsia shu-hsiang t'ung-shih she-jen* 斗 下 樞 相 通 事 舍 人)[33] and speak the [following] spell:

This evening your disciple burns incense and respectfully performs the Great Method of the Five Thunders in order to infuse [its might] into the divine seals to be jointly used to employ the [divine] emissaries. I do not dare to act alone on my own authority. I reverently submit my request to the Commissioner at the Dipper and Secretarial Receptionist, to the Commissioners of the Waterways[34] at the right and left sides, to the Lads of Jade-Perfection (*yü-wan t'ung-tzu* 玉 完 童 子), to the Lower Memorial Processors (*tsou-shih [kuan]* 奏 事 [官]) of the Five Thunders who attend to the amulets and to the Palace Guards (*kung-ts'ao* 功 曹), who are in charge today.[35] I reverently submit my wish that you all descend riding your carriages as to receive and be pleased with my ample rites [of veneration].

---

31 Or translate: "transmit initiation and ordination".
32 *Tz'u-tien* (祠 典) has the meaning of *ssu-tian* (祀 典).
33 Hucker, p. 435, nr. 5431; p. 556, nr. 7507. The divine title is possibly a mistake for *tou-chung shu-hsiang* (斗 中 樞 相), see the pantheon TT 1220: 56.6a.
34 Hucker, p.542, nr. 7282.
35 Hucker, p.527, nr. 7042; p.297, nr. 3489.

Again you salute three times, pour wine, burn incense and use **(12a)** [for the sacrifice] twelve pieces of gold currency and twelve cloud-horses (*yün-ma* 雲 馬). The left hand and the right hand form twisting [the fingers] the *mudrâ* pattern-*ch'en* (辰 文). You speak in silent meditation the [following] formula:

Heaven, help me to get the wood on a *ping* 丙 [-day].[36] Radiance of thunder, transform life. Pivot of the dipper (*tou* 斗), let your magic might descend. Change and let work the five phases (*wu-hsing* 五 行). The eight trigrams (*pa-kua* 八 卦) may be magically penetrating all around. The nine provinces may be in a state of awe. The Water Department may be the controlling agent, and the fire monsters must dissolve their outer forms. I wish to have the divine breaths descend and floating spread out everywhere in the nine [spheres] of purity. Act urgently as this is the law and order.

Having spoken the spell you concentrate on all the deities that descend and let the perfect breaths come close to you. You inhale [the perfect breaths and] lead them into the seal. You take [the seal] then back to a quiet room and must not allow women, chicken and dogs to see it. Do use [the seal] in accordance with the rules.

Spell for taking out the seal (*ch'u-yin chou* 出 印 梘): Fire bells in the palace, your scorching brightness combines and penetrates [all around]. Five Thunders of the five regions, let your radiance descend and arouse the wind. Might of heaven, terrify the four regions, and the evil forces will extinguish their tracks. Act urgently as this is the law and order.

(12b) Spell for using the seal (*yung-yin chou* 用 印 梘): August emperors of heaven do issue your orders to behead the wicked spirits and extinguish the cruel ones. The divine seal is stamped down right away, and wind and fire may thus become urgent and pressing. Act most urgently, as this is the law and order of the superior god-emperors.

Having spoken the spell you take the breath of thunder (*lei-ch'i* 雷 氣) and blow it onto the seal. You concentrate your vision on the radiance of fire that extends over a distance of ten thousand feet. Above, it links up with the pivot of the *tou*-constellation (斗), and down here [on earth], it flows into the pattern (*wen* 文) of the seal. You concentrate on the radiance of the lightning of the Five Thunders that shakes and echoes in heaven and on earth. The Heaven of the Divine

---

36 See above p. 10b, which says that the wood for a seal has to be collected on special days.

Empyrean (shen-hsiao t'ien 神宵天) will be half-blue and half-red. You inhale these breaths into your mouth and blow them onto the seal. [37]

Spell for taking back the seal (ju-yin chou 入印咒): Divine seal do return to the sun [-direction] (sun巽). Jade lads do return to your [heavenly] palaces. All you functionaries do guard your positions. When you receive my summoning, you [must] follow suit. Act most urgently as this is law and order.

After speaking the spell, you wrap up the seal and let the Thunder officials and saints control it as before. [38] If you are not performing rituals, you must not take out and air the seal.

(13a) The Locations and Departments of the Five Thunders (wu-lei suo-pu 五雷所部)

The Fire Master says: In general, there are five [sorts of] thunders, namely the Heaven Thunder (t'ien-lei 天雷), the Divine Thunder (shen-lei 神雷), the Dragon Thunder (lung-lei 龍雷), the Water Thunder (shuei-lei 水雷) and the Altar of Earth Thunder (she-ling lei 社令雷).

A commentary in small print says: another name is Thunder of the Wicked-Spirits (yao-lei 妖雷). The Thunder of the Wicked-Spirits does not accept commands from the god-emperors and therefore its appellation is "wicked spirits".

What [the thunders] rule over is not at all the same, and their departments are to be distinguished. The men who study Tao and their disciples who receive the ritual methods all attain the means they can rely upon, and so they can issue memoranda, prayers and requests. However, in the case that one does not know these methods (fang 方), one exerts the mind in vain.

As to the Heaven Thunder (t'ien-lei 天雷), hundreds of officials and thousands of generals support above [in heaven] the Jade Emperor (yü-ti 玉帝) and below [on earth] they control yin and yang (陰陽). Their might and virtue weigh

---

37 Compare the paragraph shu-fu (書符) On Writing Amulets, in TT 1220: 69.14a-16a (Wang shih-ch'en ch'i-tao pa-tuan chin 王侍宸祈禱八段錦) that echoes such statements.
38 This speaks of those divinities that the ritual addresses and involves. The formulation is opaque but we also find the formulation on TT 1220: 56.41b, where it should have the same meaning and application.

extremely heavy. At the turn of a *kalpa* the superior god-emperors dispatch by decree this thunder to descend down to the world of man in order to re-open heaven and re-examine earth, to drum and agitate *ch'ien* and *k'un* (乾 坤), to install the sun and get the moon [into position]. [The Heaven Thunder] is exceedingly honourable, which is quite beyond any words.

In case that the country [experiences] excessive heat during a few years in a row and famine and devastation rage throughout the world, it is appropriate then that all the kings in the country expose *tz'u*-texts (詞). They memorialise and make [the situation] known to the heavens and extend the information to all the [Thunder] officials as to implore them to send down [the Heaven Thunder] to help and save the world. Then it is that you can **(13b)** effectively use altars and rituals in accordance with the standards, but you must not have any erratic intentions.

The Divine Thunder (*shen-lei* 神 雷) also has hundreds of officials and thousands of generals who reside in the centre of the three realms (*san-chieh* 三 界). They all are stationed (*t'un-chu* 屯 駐) in accordance with the seasons and on behalf of heaven they operate and exert their transforming influences. In one year and within the four seasons [the Divine Thunder] issues the paroles (*fa-hao* 發 號) and dispatches orders (*shih-ling* 施 令) as to spread evenly the rain and the moisture. In case that [the people in] the lower regions were neither loyal nor pious, neither humanitarian nor faithful (*chung/hsiao/jen/i* 忠 孝 仁 義) [39] and [either] in their former lives [or] in their present time harmed all the creatures in hideous ways and unjustly amassed plenty of properties, the Three Officials (*san-kuan* 三 官) then hand in [appropriate] reports to the higher [institutions in heaven] and have the [respective] names registered in the Files of the Wicked (*e-pu* 惡 簿). The superior god-emperors order then the Divine Thunder to crusade (*fa* 伐) against [the guilty ones]. Today perhaps, when wild winds and heavy rain occur and shaking sounds of thunder punish and kill men and creatures, this is just such an event. [40] If you desire to activate the Divine Thunder, you must send up a report to the Three Monitoring Offices (*san-ssu* 三 司) and let memorials soar up to the nine pure [heavens] (*chiu-ch'ing* 九 清). Then it is that you can employ the Divine Thunder.

As to the Dragon Thunder (*lung-lei* 龍 雷), the superior god-emperors conferred upon it the Dragon Palaces (*lung-kung* 龍 宮) with ten thousand generals and

---

39 This is the traditional set of Confucian ideals and virtues.
40 This proves the operations of the Divine Thunder.

thousands of troops that support the Dragon Lord (龍君) who with his might and virtue preserves and protects the scriptures of the immortals (*hsien-ching* 仙經). In general, the Dragon Palaces in the sea store ten thousand chapters of scriptures of the immortals and immeasurable extraordinary treasures, which were also given by the heavenly emperors. The [Dragon] Thunder protects them. The Dragon Thunder rules over **(14a)** the help [that is required] for a whole region when droughts and inundations occur. If you desire to activate the [Dragon] Thunder, you send memorials that soar up to all the [spirit] officials and let them inform the superior god-emperors. [The superior god-emperors] hand down to the Chef of Dragon Thunder a warrant to proceed speedily and help. Definitely, there will be a response. Its winds are smooth and harmonious and its rain is subtle and widely spread, which is such an event.

As to the Water Thunder (*shuei-lei* 水雷), it is located at the Department of the Water Officials of the Lower Origin (*Hsia-yüan shui-kuan* 下元水官). The superior god-emperors conferred upon them the order to punish and behead the water monsters (*shui-yao* 水妖), to reward merits and crusade against faulty behaviour. Belonging to the ranks of the divinities, they also preside over the help for a whole region where disaster and drought occur. As to the officials and generals, it is all the same as with [the officials] of the Dragon Palaces. If you activate and dispatch them, you [first] must report to the officials and memorialise to the water department (*shui-pu* 水部). But after the memorials were heard, the responses will come within a certain [short] time.

Concerning the Altar of the Earth Thunder now, in the centre of a district or hamlet there were men who due to their loyalty and righteousness reprimanded the country. There were men who were pious and brave, ardent and fierce who reprimanded the rulers who deserted their battle lines. They retired then to their homes and died with anger. Their radiant, divine and magic nature assembled and formed the [Altar of the Earth] Thunder. The [Altar of the Earth] Thunder can arrest sea-monsters and dragons, arouse wild winds and fierce rainfall at unsuitable times, uproot trees and break down woods. [Such spirit-forces] regard it as fundamental to demand blood sacrifices (*hsüeh-shih* 血食), but they can bring disaster and good luck for complete regions and their subtle affairs. **(14b)**. When people offer sacrifices at the proper time, the wind and the rain will be harmonious. In the case that sacrifices and announcements (*chi-kao* 祭告) are neglected, [such spirit forces] cause fierce rainfall and wild winds, angry thunders and violent lightning that appear continuously. Great floods harm the people, the sprouts and the harvest, and thus hurt human life and existence (*hsing-ming* 性命). The people of today, be it in one province or in one territory, may well have temples (*shen-miao* 神廟) where they present prayers and requests and [also] receive [appropriate] responses. Consequently, enshrinements for the venerated ones were made, and [the Altar of the Earth Thunder] is just a matter of this sort. Formerly, the most exalted nobles of the Chao (趙) [-family] engaged in military encounters, and this exactly was the sort of event [that led to the instalment of such Altars of the Earth Thunder]. [41]

The gentlemen who study perfection and respectfully venerate *Tao* receive such oral instructions (*k'ou-chüeh* 口訣) that they become able to dispatch and agitate the [Altar of the Earth] Thunder in order to save from droughts that afflict [locations] one hundred miles afar, and they also save from disasters that may occur in just one single hamlet (*i* 邑). In general, when you urgently employ the Altar of the Earth Thunder, you must present memorials to the city god (*ch'eng-huang* 城隍) and the wicked and fierce [spirits]. The ritual sacrifices order the city god to superintend the [Altar of Earth] Thunder. Thereafter you install [the Thunder] altar and the rituals can begin. You can [use this method] only to save from droughts and to procure water, but the ritual must not be deliberated in reckless ways."

---

41 The text says: „in Chao *luan*-birds and *feng*-phoenices (鸞鳳) set their battleaxes to use and fought", see *P'ei-wen yün-fu* 2263/2. I suppose that the text refers to the generals of the Chao (趙) family that rivalled and opposed the generals who finally engineered the success of the Ch'in (秦) –armies. Compare M.E.Lewis: Warring States Political History, p.632, in: M.Loewe and E.L.Shaughnessy eds.: The Cambridge History of Ancient China, From the Origins of Civilisation to 221 B.C. Cambridge 1999. We notice that in Thunder Magic the Generalissimo Chao (趙元帥公明) is an important name see TT 1220: 69.5b and the preceding chapter in this book. Chao Kung-ming retired to the mountains when the Ch'in took over in China. He is believed to have attained *Tao* and became top rank in the divine military administration, see TT 1476 *Sou-shen chi* 4.10a-11b.

The following extended text speaks about a prominent divinity that already appeared in the preceding list of divine names (*Lei-t'ing shen-wei* 雷霆神位), albeit in abbreviated form. [42]

Sacrifice to the Great Divinity of Law and Order
(*Chi lü-ling ta-shen* 祭律令大神): [43]

In the Thunder Department (*lei-pu* 雷部) there resides the Great Divinity of Blazing Fire (Yen-huo ta-shen 焱火大神). The surname is Teng (鄧) and the name is Po-wen (伯文). Formerly he had joined Huang-ti (黃帝) in his victorious battle **(15a)** against Ch'ih Yu (蚩尤), [44] and was appointed General of Ho-nan. When the great divinity saw that Huang-ti ascended to heaven he quit his rank and entered Mount Wu-tang (武當山) where he practised self-cultivation for one hundred years. He could rise and descend following the breaths. He also saw that his contemporaries did not practise loyalty and filial piety. They killed and harmed [each other], encroached upon and cheated [each other], used force to take advantage against the weak. Moreover [he saw] that kings and assistants were not able to maintain order and control. Day and night, he developed the great desire to have divine thunders [at avail] [45] and to act on behalf of heaven in order to eradicate and crusade against all those bad and rebellious elements. He constantly pondered on [this situation] without cease, and the emanations of his anger rushed towards heaven. Suddenly, one day he transformed himself to have the beak of a phoenix and silver teeth. He grew red hair and had a blue body. His left hand held a thunder auger and his right hand a thunder hammer. His body grew to be tall one hundred *chang* and evolved wings in both armpits. When he opened them up they had a length of several hundred miles, and darkness covered all [below them]. Both eyes emitted two beams of fiery shining, lightning up and illuminating a stretch of one hundred miles. Hands and feet all had dragon claws. He could fly around and roam in the Great Void. He could swallow and bite animal spirits and monsters and decapitate and crusade against obnoxious dragons. He received from the supreme god-

---

42 TT 1220: 56.5a. Compare TT 1220: 124.1b-2a, see above Chapter I.
43 See F.C.Reiter: "A Preliminary Study of the Taoist Wang Wen-ch'ing (1093-1153) and his Thunder Magic (*lei-fa* 雷法). In: ZDMG 152, 174-175.
44 *Shih-chi* 1,1b (*Wu-ti pen-chi 1*, Huang Ti). Tung-hua Comp. Taipei 1970. Kwang-Chih-chang: "China on the Eve of the Historical Period", p.69. In: M.Loewe and E.L. Shaughnessy eds.: *The Cambridge History of Ancient China,* From the Origins of Civilization to 221 B.C. Cambridge 1999.
45 Or translate: "…to be a divine thunder"…

emperors the title of *Lü-ling ta-shen* (律令大神) and was officially part of the Divine Thunder (*shen-lei* 神雷). [46]

(15b) In the 5th month on the 5th day at *wu*-time (11 a.m. to 1 p.m.) this thunder [divinity] ascends to and enters the Lodge of the Fire Bell (*huo-ling chih chai* 火鈴之宅) at the Palace of the South, and on that day you can draft (*t'u* 圖) his apparition in a room for meditation. You use the blood of a goat, a fowl and a goose, five goat heads, five [sorts of] fruit of the season and pure wine to offer veneration for this great divinity during one day and one night. [47]

When the divinity descends, you can make him arouse immediately clouds and rain. Within a moment the [divinity can] make a clear sky, stop the wind, extinguish monsters, stop epidemics and discard pestilence. The divinity can swallow the demons [that cause] epidemics. At the time of the offering you write out two pieces of the Amulet of Blazing Fire (*Yen-huo fu* 焱火符) [48] and place them on the [altar] table. The next day you take them back. You can use them to control any epidemics and wicked illnesses. The mysteries [of this method] are reserved for the oral transmission.

The presentation of the Amulet of Blazing Fire (*yen-huo fu* 焱火符) follows on pp. **15b-16b** together with the Spell in Veneration of the Divine Force (*chi-shen chou* 祭神咒):

Great Divinity of Law and Order, Venerable [Divinity] of Wind and Fire with a pair of armed and huge wings, you roam and soar through heaven and earth (*ch'ien-k'un* 乾坤), behead the bewitching spirits and swallow evil forces, bind up the demons and retain plagues, cause rain to fall suddenly [or] make presently

---

46 In other words, he belonged then to the position and function of Divine Thunder.
47 Compare TT 1220: 57.16a, for a similar arrangement (*chi-fa* 祭法), contained in *Shang-ch'ing yü-shu wu-lei chen-wen* (上清玉樞五雷真文), in: TT 1220: 57.1a sq.
48 For this amulet see TT 1220: 56.15b-16b. The following spell (*Chi-shen chou* 祭神咒) summarises the divine capabilities and influences that are featured in the preceding text. Also compare e.g. TT 1220: 57.16b-17a and 61.15b-17a which gives a superb dissection of the amulet (*san-hsing* 散形), featuring its parts and elements. See *Yen-huo lü-ling Teng t'ien-chün ta-fa* (焱火律令鄧天君大法) which features Lord Teng as "Leading Marshal" (*chu-shuai* 主帥), right after the overlord of the ritual, *Chiu-t'ien lei-tsu ta-ti* 九天雷祖大帝 in: TT 1220: 80. 1asq. The chapter is dedicated to this Lord Teng and his amulets, see e.g.pp.6b-8a; pp.22b-33b. This text also contains instructions on the composition of amulets and the performance of ritual steps, for example see TT 1220: 80.4b-6b.

a clear sky. Obey my spell, obey my summoning. Come and descend to the gate of *sun* (巽 門) [49] and act urgently as this is law and order.

The Application of Thunder Breaths (*fu lei-ch'i* 服 雷 氣)

This section of the text says (**16b-17a**): The men who receive the [Thunder] rituals [50] save people from illness and distress. They cut off the wicked spirits, chase away what is evil and heal illness (*chih-ping* 治 病). [51] They need attain the method "the Thunder Lord (*lei-kung* 雷 公) invites the breaths", and then they can use the method [Application of Thunder Breaths]. After your initiation when the sound of thunder is suddenly heard, before or after the period Excited Insects (*ching-che* 驚 蟄), [52] you arrange properly an incense altar that faces the direction [of the thunder that you had heard]. (**17a**) Both of your hands form twisting the *mudrâ* Thunder Office (*lei-chü* 雷 局) [53] and with closed eyes you speak secretly the spell:

I received the manuals and rituals of the Five Thunders. Thunder and thunderclaps make mighty sounds. I internalize them to control thus my body and preserve my life (*pao-ming* 保 命). I spit them out to bind up tightly the demons and punish evil forces. Divine breaths across a distance of ten thousand *chang*, do water the flowers of my stomach. Supreme and Fierce Emissary of Law and Order (*lü-ling* 律 令), Silver Teeth, do act urgently as this is law and order.

After speaking the spell, you wait for the flash of lightning and the sound of thunder to occur. You look out for the direction and stare straight that way. You breathe in the breaths [of lightning and thunder] and swallow them twenty five times. There is shining in hiding (*yin-yin ming-ming* 隱 隱 明 明). When the radiance comes down you swallow most hastily these breaths, which shall have the [desired] result.

---

49 The divinity enters at this point the ritual area.
50 Alternartively say "the rituals".
51 Or they „heal illness".
52 See H.A.Giles: *A Chinese-English Dictionary*, p.26 ("March"). (Rpr. 1972 Taipei).
53 Compare F.C.Reiter: "The Discourse on the Thunders", p. 217, in: JRAS 14/3.

Writing out Amulets to Cure Illness, Raise Dragons and Produce Rainfall (*shu-fu chih-ping ch'i-lung hsing-yü* 書符治病起龍行雨)

The following section **(17a-18a)** starts out with a quotation of the Fire Master who says:

The great methods that were handed down by the Court of the Emissaries of the Five Thunders (*Wu-lei shih-yüan* 五雷使院) [enable to] write out amulets, cure illness, raise dragons, produce rainfall, drive back floods and beg for clear skies. There are ritual instructions (*fa-chüeh* 法訣) for all of this, which must not be used in unauthorised ways. In case that there are offences **(17b)** [the ritual] will not have any divine force (*pu-shen* 不神).

If you have to write out an amulet, you prepare the ritual performance (*fa-i* 法儀) and set up an altar. You twist [the fingers to show] the *mudrâ* The Divine Empyrean Calls out the Summons (*shen-hsiao hu-chao chüeh* 神宵呼召訣). You perform the [ritual] steps (*kang-pu* 罡步) in the direction of Southeast and the *sun*- (巽) window, present incense and summon the divine generals (*shen-chiang* 神將) to attend to the matter of concern. You start to memorialise to the superior god-emperors, to the Court of the Emissaries of the Five Thunders, to the Three Monitoring Offices of Thunder and Thunderclaps (*lei-t'ing san-ssu* 雷霆三司) and [finally to] the Chef Stuarts of the Five Thunders (*wu-lei chu-tsai* 五雷主宰). After this procedure you form twisting [your fingers] the *mudrâ* Transform to be Divine (*pien-shen chüeh* 變神訣) and so transform yourself (*hua-shen* 化身) to be the Emissary of the Five Thunders.[54] You walk the steps Crack the Earth and Summon the Thunders (*p'o-ti chao-lei kang* 破地召雷罡).[55] When you reach the front side of the altar, you apply the *mudrâ* Thunder Office (*lei-chü* 雷局) and speak in meditation the spell:

Divine Constellation of the Jade Pivot, receive the decree for the Perfect King to crack the earth and summon the thunders, to punish and crusade against the cruel and wicked forces. Emissaries of the Five Thunders, hurriedly come up to the site of the altar and complying with my spell and decree start to chase away

---

54 Compare F.C.Reiter: "A Preliminary Study of the Taoist Wang Wen-ch'ing (1093-1153)", in: ZDMG 152/1, p. 172, translating "Assembling the Divine Force" (*Lien-shen* 鍊神); see TT 1220: 124.1b-2a. Again, the divine force is being created within the own body. In this sense, there is a clear identiy of man and god. The concept of "spirit possession" (shamanism) most certainly does not apply.
55 Compare TT 1166 *Fa-hai i-chu* 6.23a-23b.

what is unlucky. Report to the five [thunder] departments that they let greatly descend their divine radiance. [The divine radiance] may flow into my brush and ink slab to write out the heavenly tracts [56] (t'ien-chang 天章) in seal script. The water monsters may perish and any evil ways [57] (hsieh-tao 邪道) may get scattered and vanish. Do act urgently as this is the law and order of the god-emperors of Thunder and thunderclaps (lei-t'ing ti-chün 雷霆帝君).

You speak the spell five times and concentrate [your vision] on the Emissaries of the Five Thunders, the Supreme Counsellors at the Jade Pivot, the P'eng-lai-Emissaries (蓬萊使者) of the god-emperor of the North (pei-ti 北帝). (18a) The thunder carriages together with the thunder divinities of the five departments, the officials, generals, servants and troops altogether descend down from heaven, come forth from the earth and hurry up to patrol. You reflect upon and visualize that the generals of the Five Thunders and their servants all transform their bodies (hua-shen 化身) and appear to be thick mist that together with the fire of the thunderclaps spread all around your brush, your ink slab, the red ink and also in the water and on the paper. After that, you grind your teeth (k'ou-ch'ih 叩齒) five times, [58] take the brush and shout out very loudly five times. You inhale the breaths five times (wu-k'ou 五口) and blow them onto the brush and the ink slab. You concentrate [your vision] on the radiance of fire that [shines] over a distance of ten thousand feet and combines a bright shining that projects on the person (she-jen 射人) who is responsible and commissioned to write the amulet.

A drawing of the Amulet of the Comprehensive Assistance of the Five Thunders (wu-lei tsung-she fu 五雷總攝符) follows. The amulet shows the character for thunder (lei 雷) that a curved line with five smaller circles on its inner periphery encircles. (18b) We learn "that the amulet on the right hand side represents the character thunder (lei 雷). After its completion the character or the drawing was enlarged with five [small] rings (ch'üan 圈) called Thunder One, Thunder Two, Thunder Three, Thunder Four and Thunder Five. [59] Twisting the hand to form the mudrâ [pattern-] sun (巽) [you can use the amulet] to control the breaths of epidemics, illness and the wounds that were caused by cold and poison or any other wicked illness."

---

56 Or: „memorials for the heavens"
57 „evil ways" is a rather opaque phrase which may point to competing ritual or exorcist specialists.
58 This is an established method, see F.C.Reiter: Der Perlenbeutel aus den Drei Höhlen, pp. 166-172, referring to TT 1139 San-tung chu-nang 10. 1a-11b.
59 See F.C.Reiter: "The Discourse on the Thunders", p.213.

Now we get the Amulet of the Five Locks (*wu-so fu* 五鎖符). The graphic design shows the character for the word well (*ching* 井) within a circle.

The spell [for this amulet] says: Thunder Deity of Heavenly Fire, Thunder Deity of Earthly Fire and Five Thunders let your magic might descend to lock up the demons and bar evil spirits.

The explanation states: The spell has to be spoken when you write the character *ching* (井). After this, you turn the brush around and smear (*t'u* 涂) seven times [seven circles on the character *ching* 井].[60] The same time you speak in your mind the spell: **(19a)** At the first turn [of my brush], the six spirits (*liu-shen* 六神) hide away. At the second turn [of my brush], the four killing forces (*ssu-sha* 四煞) perish. At the third turn [of my brush], the *k'uei*-star (*k'uei-kang* 魁罡)[61] is agitated. At the fourth turn [of my brush], the fire of thunder rushes upwards. At the fifth turn [of my brush], the thunderclaps charge forth. At the sixth turn [of my brush] the mountain demons die. At the seventh turn [of my brush], what rebels against heaven and does not have *Tao* is collected and arrested to have the heads chopped off and the feet severed. The fifteen types of deities and demons that are not correct and cause misfortune rush altogether to surrender to my Five Thunders. They are not allowed to agitate and move. Act urgently as this is law and order.

The instruction says: After the spell was spoken, you take off the brush [with a quick movement of the hand] towards the direction of Southeast and exhale [noisily] the thunder breaths that enter the amulet. You concentrate on the sound of thunder that shakes the earth and on the radiance of fire within the amulet that reaches as far as ten thousand feet, and you bless then dispatch [of the amulet].

Now we find the *mudrâs* that go with the Amulet of the Five Locks (*wu-so fu-chüeh* 五鎖符訣): [When] heaven and earth are shaken by five sounds [of thunder] you write the character "well" (*ching* 井) and concentrate [your vision] on the fire of thunder and thunderclaps that enter the character *ching* (井). You form the *mudrâ* [pattern-] *wu* (午), write the character *ching* (井) a second time and meditate on the Thunder Deity of the Fire of Heaven (*t'ien-huo lei-shen* 天火雷

---

60 I understand that in the end the amulet or character would be hidden under these circles in black ink.
61 The first star of the Big Dipper, compare F.C.Reiter: "The Discourse on the Thunders", p. 220.

神). After this, you form the *mudrâ* [pattern-] *ch'ou* (丑) and meditate on the Thunder Deity of the Fire of Earth (*ti-huo lei-shen* 地 火 雷 神), and then you meditate on the Thunder-Fire One, the Thunder-Fire Two, the Thunder-Fire Three, the Thunder-Fire Four **(19b)** and the Thunder-Fire Five. When you pass the twelve branches (*chih* 支) [pointing them out on your palm], the sequence starts out at the position of *tzu* (子) to pass [the position of] *wu* (午) and the twelve branches [altogether]. When you reach [again] the position of *tzu* (子), you form the [*mudrâ*] Thunder Office (*lei-chü* 雷 局). You meditate then on the Five Thunders that send down their divine might, lock up the demons and bar the evil spirits. Now, you proceed as above and recite in meditation the spell on the seven circles [drawn by the writing brush] (*ch'i-chuan chou* 七 轉 咒). At the first turn [of the brush], you form the *mudrâ* [pattern-] *mao* (卯).[62] At the second turn [of the brush], you form the *mudrâ* [pattern-] *yin* (寅). At the third turn [of the brush] you form the *mudrâ* [-pattern] *ch'ou* 丑. At the fourth turn [of the brush], the middle finger (*chung-chih* 中 指) points to the central line on the palm (*chung-wen* 中 文). At the fifth turn, the fourth finger points to the upper line [on the palm]. At the sixth turn, the fourth finger points to the central pattern [on the palm]. At the seventh turn, you form the *mudrâ* [pattern-] *tzu* (子)." [63] The concluding statement in this paragraph says that this [spell and method] rely on the above method of smearing circles of black ink on [the amulet] with the writing brush that finally must be taken off [with a quick movement of the hand] towards the direction of Southeast.

The result of the action is show on **p.19b**, which shows a fat and round black dot with a slim tail pointing upwards beyond its left upper part. This indicates the movement of the brush that was quickly taken off. The black dot hides the amulet.

The following amulet has the title Amulet of the Green Dragon of the Breath of Life (*sheng-ch'i ch'ing-lung fu* 生 氣 青 龍 符) that is good for gaining prosperity and keeping monsters under control (*chao-ts'ai chen-kuai* 招 財 鎮 怪). The description says: First inhale the breath of life of the East and let it enter the palace of the liver. You blow then the breath onto the tip of the brush that

---

62 This most certainly means that the individual positions of the branches of earth on the palm are being pointed out (*tien* 點) or counted, which I call a *mudrâ*.
63 See F.C.Reiter: "The Discourse on the Thunders", pp. 212-221, in: JRAS 14/3. These chronological and calendar elements are most important for Thunder Magic, at least in the perception of Wang Wen-ch'ing (王 文 卿). Obviously, any time some position on the palm is being pointed to or pressed upon, we have a specific pattern *mudrâ*. In some statements, the word *wen* "pattern" (文) obviously means "lines".

together with your eyes shines radiantly, **(20a)** and write the amulet. You form the *mudrâ* [pattern-] *chen* (震), grasp the breath of life and concentrate [your vision] on the green dragon that riding the breath enters the amulet.

The amulet is shown in a dispersed form (*san-hsing* 散 形) that is used for learning and memorizing the composition and the internal meaning of the drawing. The amulet itself must be done in a fast and concentrated action that results in the assembled or complete form (*chü-hsing* 聚 形) of the amulet. We see on **p.20a** the ideal form of the amulet. The meaning of the individual parts of the amulet is stated as follows: Shaking thunders rise fast, fire wheels display their abundance. Three-and-Five, the Iron Face, hurriedly rides on strong smelling smoke. Dragons speedily hurry on like lightning, riding rapidly on cloud carriages. They call in prosperity (*chao-ts'ai* 招 財), keep guard on residences and assist to destroy the sources of wicked influences.[64]

Another amulet follows. It is called the Amulet of the Soaring Sword that Arrests Dragons (*fei-chien chuo-lung fu* 飛 劍 捉 龍 符). The commentary explains that the amulet keeps guard on residences and sites (*chai-lung* 宅 龍)[65] and brings security to men and creatures. It severs and subdues demons and all those spirit forces that kill. P. **20b** shows a few small graphic designs that altogether constitute the amulet. The explanations for each single part of the drawing are telling: The military detachment on the left hand side and the military detachment on the right hand side, both of them shake roaring the three realms. Lord of Thunder and Mother of Lightning, arrest quickly the wicked spirits. Generals of the Thunder Lord of the Green God-Emperor; generals of the Thunder Lord of the Red God-Emperor; generals of the Thunder Lord of the White God-Emperor; generals of the Thunder Lord of the Black God-Emperor; generals of the Thunder Lord of the Yellow God-Emperor, do rise quickly, do rise quickly as to arrest the wicked and bewitching dragons.[66]

P. **21a** contains the Amulet Raise the Dragon and Cause Rainfall (*ch'i-lung chih-yü fu* 起 龍 致 雨 符).[67] The decree and spell for the amulet say: Black clouds,

---

64 Due to small printing, some of the characters are rather difficult to identify. The translation is tentative.
65 "dragon" could refer to promontories or hilly sites where houses are built, in accordance with the diction of *feng-shui* (風 水). For example, see S.D.R. Feuchtwang: An Anthropological Analysis of Chinese Geomancy, p. 15. Vientiane 1974.
66 The colours represent the totality of the cosmos and its positive spirit forces. For the generals, see the pantheon TT 1220: 56.9b.
67 An alternative name is indicated in small print that is not clearly readable.

which is the auspicious sign may ascend. T'ai-i (太 乙) may support [the rise of] the morning. Jade dawn and divine force of the cavern let the *yin*-[darkness] (陰) sink down to give reign to the dragons. The iron plate [with the amulet] was just now exposed. [68] Thunder masters (*lei-shih* 雷 師) prop up [the shafts of the Thunder] carriage: Grand Senior T'ai-shang (*T'ai-shang hao-weng* 太 上 浩 瓮), most urgently receive [this] order from the Lord of Origin (*Yüan-chün* 元 君).

The subsequent explanation gives the following information: When in commanderies and cities excessive droughts lasted a long time and any prayers for rain had not been answered, you then search out famous mountains **(21b)** and grotto-caves (*tung-hsüeh* 洞 穴) that have sites with dragon-pools. You suitably set up an altar in accordance with the rules and recite the spells. [69] Consequently, you take a [square] iron plate that has a length of nine *ts'un* and a width of three *ts'un* and use red ink to write the precious amulet. You speak in meditation the spell forty nine times, and then you throw the plate with the memorial [and amulet] down into the pool. You dispatch all the [spirit] functionaries, officials and generals who pledge then to receive the city god and the god of the earth (*ch'eng-huang she-ling* 城 隍 社 令) to raise together the thunders and cause rainfall. They will come down right in time. First, you must prepare and get in [proper] form the memorials and petitions, and then everything is just fine.

You use unwrought iron in order to cast the [iron] plate that is thick one *ts'un* and broad seven *ts'un* for the single ritual. The plate has a length of one *ch'ih* and two *ts'un*. When there is the occasion for prayers and requests, you also write out four pieces of the Amulet Raise the Dragon and Cause Rainfall [*ch'i-lung chih-yü fu* 起 龍 致 雨 符]. You put them together with an Amulet of Floating Mercury (*liu-tan fu* 流 丹 符) into the four jars (*weng* 瓮) that you place on the four corners of the altar.

Now we see a drawing of the just mentioned Amulet of Floating Mercury. **(22a)** The following description says that the amulet uses an iron leaf (*t'ieh-yeh* 鐵 葉) that is long nine *ts'un* and broad three *ts'un*. You engrave the amulet using red [colour]. You do this a second time to fill in [the red colour] where it was deficient.

---

68 This most certainly refers to the practice known as the „exposure of dragon tablets" (*t'ou lung-chien p'in* 投 龍 簡 品). See F.C.Reiter: Der Perlenbeutel aus den Drei Höhlen pp. 39-41. Compare TT 1139 *San-tung chu-nang* 2.8a-12a.
69 This is the preceding text.

The Amulet of the Great Divinity of Blazing Fire (*yen-huo ta-shen fu* 焱 火 大 神 符) is the following theme. [70] The amulet consists of a number of graphs for which we get some explanations: The left eye is the sun. The right eye is the moon. Open the eyes, and the shining of fire shines over a distance of ten thousand *chang*, illuminates heaven and earth, and extinguishes evil demons. Heaven is round and earth is square. There are six rules, nine paragraphs and eight trigrams (*pa-kua* 八 卦). [71] Emissary of Savage Thunder of the East, do most speedily rise up. Emissaries of the Five Directions just do the same. **(22b)** The supreme god-emperors (*shang-ti* 上 帝) have the decree to sweep clean the nine provinces. Where the response is not compliant, [the perpetrator] is to be fixed below the five holy mountains. Act most urgently as this is law and order. Fire of thunder, do burn and destroy. Demons will be terrified and divinities will be grieving. On the right side you blaze [with flames] [72] Mount K´un-ch´iu (崑 丘) – and the four rivers (*ssu-tse* 四 澤) stop their flow. On the left side you blaze [with flames] the five holy mountains – and the stars fall down from heaven. You drag the heaven and pull the earth. Your hands rout the clouds of fire. Soaring high you search the cosmos (*liu-ho* 六 合) and let lightning stream through the great void. Dragons find themselves tied up. Demons and divinities extinguish their traces. Fire chariots approach hurriedly across [a distance of] ten thousand *chang*. The heat [of the sun] (*ping-ting [huo]* 丙 丁 [火]) destroys any doubts, burns demons (*fan-kuei* 梵 鬼) and extinguishes disasters, [73] urgently, most urgently!

(23a) The following decree and spell say: Generals and emissaries of the Three Monitoring Offices, today you attend to this amulet (*chih-fu* 直 符) and following me urge the emissaries to accept respectfully and obey the heavenly amulet (*t´ien-fu* 天 符) in order to collect and arrest the demoniac thieves, to push back illness and disperse annoyances. What dares to rebel against the command,

---

70 Despite the grand title, the amulet and its description are abbreviations of the Amulet of Blazing Fire in: TT 1220: 61.15a-17b. Concerning this amulet, see F.C.Reiter: "The Management of Nature: Convictions and Means in Daoist Thunder Magic (Daojiao leifa)", in: AAS 29, pp. 203-205, in: F.C.Reiter ed.: Purposes, Means and Convictions in Daoism. There are a few minor variants in the translation. Concerning the divinity Great Divinity of Blazing Fire, also see TT 1220: 56.6a of the pantheon.

71 "Six rules" may refer to the six upper musical pitchpipes of ancient music, and "nine paragraphs" may stand for the nine branches of mathematics, which together with the eight trigrams describes the totality of the cosmic order using antique and formal expressions.

72 I take the character *shan* (煽) to stand for *shan* (煽).

73 The three characters after *ping-ting* (丙 丁) are almost unintelligible. This is a tentative translation.

Writing out Amulets to Cure Illness, Raise Dragons and Produce Rainfall   95

thunder axes may speedily execute. The superior god-emperors issue this command to drive away and expel [all molestations].

Having spoken the spell you open [the hand with the *mudrâ*] Thunder Office (*lei-chü* 雷 局) and let it enter (*ju* 入) the amulet. You burn [spirit] money, [spirit] horses and the memorial (*cha-tzu* 劄 子). You dispatch the divine generals and palace guards **(23b)** that go along with the command of the amulet and disperse and control. The results will immediately become evident.

We learn that this amulet shows the true form (*chen-hsing* 真 形) of the great divinity. It can immediately bring about a clear sky or rainfall, punishes and crusades against wicked forces, beheads evil spirits and expels bad influences, swallows and devours the demons of epidemics. All the divine effects [of the amulet] can hardly be reported. This is the supreme amulet in the centre of the thunder departments, which is available. Formerly, the Perfect Lords Wu [Meng] (吳 猛) and Hsü [Sun] (許 遜) and the followers of these masters all received this mystery.[74]

Now we find the Amulet that Circulates Thunders and Hits Evil Spirits (*yün-lei ta-sui fu* 運 雷 打 祟 符). **(24a)** The ritual method for the amulet [works as follows:] [The fingers of] both hands start pressing (*tien* 點)[75] [the position of] *mao* (卯) [on the palm] and continue to press the [following] positions of *ch'en* (辰), *wu* (午), *shen* (申), *hai* (亥) and *wei* (未) except [the position of] *tzu* (子). Freeze [the hands] that clutch tightly the *mudrâ* Five Thunders (*wu-lei chüeh* 五 雷 訣). You start out from the sides of both ears to move your hands [with the *mudrâs*] around in a circle, and [at the same time] you recite in meditation the Spell Five Thunders (*wu-lei chou* 五 雷 咒). You perform then the ritual paces Cracking the Earth and Summoning the Thunders (*p'o-ti chao-lei kang* 破 地 召 雷 罡),[76] shout out to summon the Five Thunders and concentrate upon their arrival. You look out for the locality where the evil forces are, take the two hands with the *mudrâ* Thunder Office (*lei-chü* 雷 局) and press them down on both ears that you strike as the Thunder drums (*lei-ku* 雷 鼓). You let both eyes flash to be the radiance of lightning (*tien-kuang* 電 光) and spit out water to be wind and rain. You release the *mudrâs* (*fa-chüeh* 發 訣) and focus [your meditation] on the rumbling of thunder and thunderclaps (*lei-t'ing pi-li* 雷 霆 霹

---

74 Concerning the two Taoists, see J.M.Boltz: A Survey of Taoist Literature, Tenth to Seventeenth Centuries, pp. 70-72. Berkeley 1987. See F.C.Reiter: "A Preliminary Study of the Taoist Wang Wen-ch'ing", p. 159-160, in: ZDMG 152/1.
75 The following positions are all earthly branches.
76 See above and TT 1166: 6.23a-23b.

霹), that roaring evolves and proceeds to decapitate and extinguish the evil and wicked spirits.

The Spell Gather at the *Kang*-[Stars] (收 罡 斗 咒) follows: All stars (*hsing-tou* 星 斗) come back to the *kang*-[position] (罡).[77] I lead my steps back to the Pure Hall (*ch'ing-t'ang* 清 堂)[78]. Divine animals support me. All deities support me [as my spirit] generals. Whatever I ("my person" *wu-shen* 吾 身) aim at, sun and moon are radiantly shining just as much. Act urgently as this is law and order.

The Establishment of Hells for Evil Demons
(*Chih-hsieh chien-yü fa* 治 邪 建 獄 法)

(24b) The descriptive explanation of the ritual[79] refers to the Fire Master who says as follows: the great method of the Five Thunders (*wu-lei ta-fa* 五 雷 大 法) especially sends down [to us] the secrets of the jade amulets that control evil forces and behead wicked spirits. In unusual circumstances they save the living beings from the dangers of heavy distress and calamities, cleanse all creatures from their dark and gloomy faults, exterminate shrines (*lu-miao* 戮 廟), expel bad forces, punish wicked spirits and cure diseases. For all of this there are specialised offices [that are in charge]. In the case that such wicked spirits and evil forces (*ching-hsieh* 精 邪) are around in the world you just take amulets and [consecrated] water (*fu-shui* 符 水) that you provide either to be swallowed or to be carried [on the belt].[80] Peace and health will come in due time.

In case that those wicked forces have an outer shape and substance, they soar in the void or walk on earth. Some wicked divinities in the water and on earth cause calamities. Water monsters (*shui-yao* 水 妖) cause misfortune (*nieh* 孽), and when they all do not yield to the control by amulets, there is the method of establishing hells (*chien-yü fa* 建 獄 法). You command the Three Monitoring Offices (*san-ssu* 三 司) to arrest, bind up and transfer [such malignant elements] to the hell, where they are checked and brought under control. [In this way] you

---

77 This should refer to „...the bowl of the constellation the Big Dipper".
78 This may be the heavenly sphere that is visited in meditation as to adopt the radiant might of the central cosmic sphere that stars define.
79 Alternatively translate simply: "method". This theme is connected with rituals of healing, see, for example, TT 1220: 59.1a-10b.
80 Obviously *t'un-p'ei* (吞 佩) are two different modes of application, otherwise the translation "amulet water" would have been fine.

eliminate forever their very roots and origin. After establishing a hell, you can use it according to the rules.

Now we find the title: Hell of thunder and lightning that check and control (*lei-tien k'ao-chih yü* 雷 電 考 治 獄). The square drawing of the hell is rather poor. However, we get the following information **(25a)**: In the three realms (*san-chieh* 三 界), the hell bars and arrests all those, who belong to [spirit] officers and generals, managers of sacrifices,[81] to the immortal officers of the lower echelons and [other] officials who do not have *Tao* and belong to the demons and divinities who [wrongly] adopt personal names and attach themselves to people to commit any sorts of faults and insult and distort Taoist rituals (*tao-fa* 道 法) and do not surrender to heavenly amulets (*t'ien-fu* 天 符). You establish this hell [for them].

The following hell is called: Hell of the comets with the radiance of fire (*huo-kuang liu-hsing yü* 火 光 流 星 獄). We learn that this hell is made to put away and arrest scaly dragons, poisonous scorpions and any sorts of water monsters. When they have fully attained their capacity, they rise up to **(25b)** reach [the rank of] *yin*-officials (*yin-kuan* 陰 官) in the Water Department (*shui-fu* 水 府). Without any reason they harm people, [spoil] sprouts and the harvest, let water shoot forth, swell rivers, capsize passenger boats, swallow and devour living beings, and being without *Tao* they cause any disasters. You use this [hell] to bar and control [them].

The following hell is called: Hell of the poisonous and harmful metal spirits (*chin-ching tu-hai chih yü* 金 精 毒 害 之 獄). This hell puts away and bars the mountain spirits and wood deities (*shan-ching mu-shen* 山 精 木 神) in the world, the wicked spirits that have fully attained the capacity of the five elements. Only these monsters (*ching-kuai* 精 怪) are wicked spirits. They have outer shapes and figures and attach themselves to the physical form of the human body. Sometimes [such elements] are in hiding and sometimes they become visible. You use this hell for all of them.

The hell is in charge of the trouble caused by the Nine Wells (*chiu-ch'üan k'u-nao chih yü* 九 泉 苦 惱 之 獄).[82] This hell is said **(26a)** to put away and bar epidemics and any sorts of demons (*mo-kuei* 魔 鬼) that cause trouble for the

---

81 See *Hucker* nrs. 1015,6650.
82 The name Nine Wells stands for the Hades or for death, see H.A.Giles: A Chinese-English Dictionary nr.2263. Rpr. Taipei 1972.

condition and outer appearance of temples and harm the natural disposition of people. [The hell puts away and bars] demons and divinities (*kuei-shen* 鬼 神) that in the world operate as savage robbers and thieves who do not comply with Great *Tao*, harm the country and hurt the people. This hell bars them all.

The hell of the gloomy platform-tower and the long night (*yu-t'ai ch'ang-yeh chih yü* 幽 臺 長 夜 之 獄) **(26b)** puts away and bars what is not connected with correct sacrifices (*cheng-ssu* 正 祀). When clan members formerly were not in accord with *Tao,* they conceal themselves in corpses. As itinerant demons and wicked spirits (*k'o-kuei hsieh-shen* 客 鬼 邪 神), they are dependent on outer [physical] forms. [In this way] they enter dreams, throw pebbles and strike with stones, rob and occupy men and creatures, debauch wives and usurp rooms and porches of the people. The five [social] relationships (*wu-tao* 五 道) and everything are no longer in good order. These false divinities and pretentious immortals (*wei-shen cha-hsien* 偽 神 詐 仙) eternally cut off the roots and sources [of the people].

As a response, you set up the hell. You deliberate in detail the big and small arrangements and principles, send soaring petitions to the Three Monitoring Offices (*san-ssu* 三 司) and memorialise vis-à-vis the supreme god-emperors a complete report that explains the source of the wicked causes and where their presence can be spotted. You also state, which hell you established at a southeastern location.[83] You then beg that the Five Thunders of the Three Monitoring Offices (*san-ssu wu-lei* 三 司 五 雷) may be sent down to a visit on earth, first of all to put away and arrest the lewd and malignant forces of any quality and to rush them down into the hell where they are barred and controlled.

When you have written the report in accordance with the [proper] style, you approach then the family where the wicked person lives. You choose and cleanse a room in the southeastern [part of the house]; using incense and water to wash away any faults. You take the ashes of the incense burner and establish the hell [laying out the ashes on the ground] according with the model.[84]

Every hell [of this type] has a diameter of nine *ts'un*. You use one branch of a peach tree with a length of three *ch'ih* and fasten red silk [streamers] on its upper

---

83 This refers to the notorious *sun*-direction (巽).
84 The ashes are used to design or lay out the hell on the floor.

Altars and Prayers for Rain or Clear Skies 99

end. Each **(27a)** of them is long five *ts'un*. You stick [the branch] into the centre of the gate of every hell and attach its name [on the branch].

After this, you burn incense and invite the Chef Managers (*chu-tien* 主 典) who investigate vicious demons and revise and correct [the archives of] the good ones and the bad ones. You invite the Chef Stewarts (*chu-tsai* 主 宰) of the Five Thunders [85] who are in control of the hells, the Emissaries who are in control of the ordinances and investigate by torture (*tien-lu k'ao-chin shih che* 典 錄 拷 禁 使 者), the Emissaries who bind up the *hun*-souls and deport them under supervision (*chuan-hun chien-sung shih-che* 縛 魂 監 送 使 者), the Emissaries of Law and Order (律 令 使 者), the Emissaries who transmit the orders (*ch'uan-ling shih-che* 傳 令 使 者) and the Thirty Six stalwarts of the thunder-drums (*san-shih-liu lei-ku li-shih* 三 十 六 雷 鼓 力 士), [86] and you implore them all to descend and form a guard for the hell to lock up its gates.

It does not take very long and their apparitions become visible. After their appearances became visible, they measure the weight of the crimes, whether they have to execute [the culprits] or imprison them. Perhaps they send them to the Office of Thunder and Thunderclaps of the Emperor of the North (*Bei-ti lei-t'ing ssu* 北 帝 雷 霆 司) and the *P'eng-lai* Office (蓬 萊 司), where their cases will be judged according with the law. Only the Court of the Jade Pivot (*yü-shu yüan* 玉 樞 院) does not correct crimes (*chih-tsui* 治 罪).

---

85   See *Hucker*, p.514, nrs.6515, 6809.
86   For the emissaries see, TT 1220: 56.8b of the pantheon. For the thirty-six stalwarts, see p.9a. Many of the divinities in the pantheon do not appear in our texts. However, the names that we can identify in the pantheon are just enough to prove that the materials in *chapter 56* are homogeneous. Concerning the method of "binding up" (arresting) souls, see TT 1220: 59.7a-7b, which involves the *mudrâ* Thunder Office (*lei-chü* 雷 局) and a spell addressing the Five Thunder Emissaries (*wu-lei shih-che* 五 雷 使 者).

Altars and Prayers for Rain or Clear Skies

The presentation first features the model for the altar that is used to pray for rain (*ch'i-yü t'an-shih* 祈 雨 壇 式). The Fire Master says that the official administrators will have to expose *tz'u*-prayers (投 詞) when droughts last long in all the territories of the country, when the provinces and commanderies lack rain.[87] The officials pray that relief and help would be brought about, and so it is that they record the prayer-texts for exposure to memorialise and make [the situation] known **(27b)** to the superior god-emperors. At the same time, they notify the Court of the Emissaries of the Five Thunders (*wu-lei shih-yüan* 五 雷 使 院) and the Three Monitoring Offices of Thunders and Thunderclaps (*lei-t'ing san-si* 雷 霆 三 司). They beg that [these institutions] especially send down instructions to direct the [spirit] officials at the lower echelons that an amnesty is granted for culprits among the people of the whole region. After this, [the local officials] implore that some specific Thunder may send down a lot of rain and moisture to save what got scorched and dried out. When the memorial dispatch for any specific thunder was made, asking that a lot of rain may quickly fall but obviously there is no response at all, [the officials] go then to a spot that corresponds above [on heaven] with the *T'ai-sui* [star] (太 歲). They take good yellow clay and build an altar with three levels.

The upper level is broad two *chang* and four *ch'ih*. The middle level is broad two *chang* and eight *ch'ih*, and the lower level is broad three *chang* and two *ch'ih*. Each of them is high one *ch'ih* and three *ts'un*. The officials place new long-necked earthen jars holding four flourishing branches at the four corners of the altar. They use a dragon pool to immerse four elixir amulets (*tan-fu* 丹 符). The incense table (*hsiang-an* 香 案) must be placed in the centre [of the altar] in order to venerate the commissioners (*shu-hsiang* 樞 相)[88] of the Three Monitoring Offices (*san-ssu* 三 司). Altogether, there are four of them. Veneration is offered to all the officials and generals of the thunder departments[89] (*lei-pu* 雷 部) at the four sides of the ground level of the altar. One employs seven priests (*dao-shih* 道 士) to stand on the altar and present the notifications [about the matter of concern]. At three points of time, namely *ch'en*- (辰), *wu*- (午) and *yu*- (酉) the

---

87 The term "official" (*kuan-ssu* 官 司) alludes to the official sphere of the administration. However, in this case the spirit official or Taoist priest should be addressed.
88 See *Hucker*, nr. 5431, p. 435.
89 Alternatively say "thunder categories".

ritual master (*fa-shih* 法師) grasps his sword. At the Window of Earth (*ti-hu* 地戶) of the altar he starts to perform the ritual paces "the soaring **(28a)** dragon brings rain" (*fei-lung chih-yü kang* 飛龍致雨罡), and so he ascends the altar.

A commentary in small print says that there is also the (alternative) name "paces to produce water and summon the thunders" (*tso-shui chao-lei kang* 作水召雷罡).

The ritual master ascends the alter and walks around it nine times, turning to the left hand side, and nine times he recites the spell "summoning the dragon to bring rain" (*chao-lung chih-yü chou* 召龍致雨咒). He says then in a loud voice: In this-or-that area, rain and moisture have failed to come in time. The *Tao* of Heaven was set for a long drought. I look up with respect to the Five Thunders and down here petition to have the service of the Emissary who Attends to the Amulet that he may quickly arouse the Three Monitoring Offices of Thunders and Thunderclaps to direct any thunder to send speedily a lot of rain in order to save what had been scorched and dried out. I have already memorialised [this problem] and made it known to the supreme god-emperors. Act most urgently and quickly without any offence against my order.

Having finished shouting aloud [these words the ritual master] descends again from the altar at its Window of Earth. It does not take longer than three days until thunder, thunderclaps and rain occur. Then again, the ritual master prepares memorials to make it known and express thanks vis-à-vis the supreme god-emperors. The ritual master arranges a *chiao*-ritual (醮),[90] returns the clay of the thunder altar and the earthen jars handing them over to the river *Ch'ang-chiang* (長江).[91] A graphic design depicts the altar, reflecting the indications in the preceding text. The picture also shows the characters Gate of Heaven (*t'ien-men* 天門) above the top level of the altar and the characters Window of Earth below the first level of the altar. **(28b)** A short line above the sketchy drawing of the altar explains that the altar table is to be installed on the top level. Below the drawing, we read where the veneration has to be offered: Below the altar at its four sides.

---

90 This serves a thanksgiving festivity.
91 This is the sole indication that could suggest the origin of these materials in a specific geographic setting which is Chiang-hsi province from where Wang Wen-ch'ing (王文卿) hails (Chien-ch'ang [建昌]district).

The next theme is the model for an altar to pray for a clear sky (*ch'i-ch'ing t'an shih* 祈晴壇式). The Fire Master says: When in the territories of the country, in provinces and cities, rain pours down continuously for [many] months in a row, [the rivers] Chiang and Ho swell, and if this development is not due to heavenly orders (*t'ian-ming* 天命) then it is due to bewitching dragons and water monsters. They cause misfortune (*nieh* 孽) to harm the people and the creatures. If the rulers of the country and the regional wardens (*shou-t'u chang-kuan* 守土長官)[92] wish to pray for a clear sky and for the rain and water to recede quickly, then it is necessary to inform (*shen* 申) the Court of the Emissaries of the Five Thunders and the Three Monitoring Offices of Thunders and Thunderclaps. The rulers of the country and the leading regional wardens altogether memorialise to the supreme god-emperors, present a *tz'u*-[complaint] to point out that it all looks like a proper heavenly disaster (*t'ien-tsai* 天災).[93] They will ask then for a basic amnesty for the culprits among the living beings, and that the rain would be put away and the floods be pushed back. As they fear that bewitching water [-spirits] (*shui-yao* 水妖) committed faults, they ask for the dispatch of the officials and generals of the Three Monitoring Offices **(29a)** and all the battalions of the Five Thunders (*wu-lei pu* 五雷部眾) to descend, put away and crusade [against such evil elements].

When on the second day there is no response at all, on the third day then you choose a high mountain or a place without any water. You take yellow clay and build at the site an altar with three levels that has the extensions of the type of altar [that has to be built to] pray for rain. However, [in this case] each level of the altar is only thick one *ch'ih* and two *ts'un*. When the construction of the altar is complete, you place the incense table on the upper level. You present then veneration for the Fierce Divinity that Carries the Wind on the Back (負風猛神), for the Great Divinity Earl of the Wind (*feng-po ta-shen* 風伯大神) and the Great Divinity of Blazing Fire (*yen-huo ta-shen* 焱火大神).[94] Veneration must be offered for the rulers of the four offices (*ssu-ssu chu* 四司主) on two sides [of the altar]. You place one empty long-necked jar in each of the four corners of the middle platform of the altar, all made of red iron ore. You put into each

---

92 Compare *Hucker*, nr. 143, p. 110.
93 Of course, the ritual presentation is the task of the Thunder specialist and priest, who is employed to perform the rituals. See below, the address is "you".
94 This is Teng Po-wen (鄧伯溫), see F.C.Reiter: "Preliminary Study of the Taoist Wang Wen-ch'ing", pp. 172-173, in: ZDMG 152/1; and the same: "The Discourse on the Thunders", pp. 222-225, in: JRAS 14/3. See the pantheon, TT 1220: 56.6a. Concerning the Fierce Divinity that Carries the Wind on the Back (*fu-feng meng-shen* 負風猛神), see TT 1220: 81.1a sq. and the preceding chapter in this book.

jar one Amulet of the True Shape of [the Great Divinity] of Blazing Fire (*yen-huo chen-hsing fu* 焱 火 真 形 符). All around on the four sides [of the altar] and on the ground level you receive respectfully the Thunder Departments (*lei-pu* 雷 部). You employ Taoist priests (*tao-shih* 道 士) as usual who invoke and display the ritual procedures. [95] The head priest (*chu-fa* 主 法) chooses three points of time, namely *yin* (寅), *wu* (午) and *hsü* (戌) to grasp his sword and perform the [ritual] paces "urge the dragons to suck up the flood" (*ch'ü-lung hsi-shui kang* 驅 龍 吸 水 罡). He ascends the altar at the Window of Earth (*ti-hu* 地 戶). [96] The priest walks forty-nine circuits around [the altar] towards the right hand side, and he recites "the spell to put away the flood and make a clear sky" (*shou-shui tso-ch'ing chou* 收 水 作 晴 咒). When the recitation of the spell has ended, [the head priest] burns an Amulet of the True Shape of [the Great Divinity] of Blazing Fire in the incense burner. There are oral instructions (*k'ou-chüeh* 口 訣) for this. Consequently, [the priest] descends **(29b)** from the altar. When right in time the wind arises, the clouds are put away and the rainfall stops, then it is that you arrange a *chiao*-festivity (醮) to thank for the [divine] mercy and send up [an appropriate] documentation. All of this must accord with the rules. After ending these procedures, you bury the clay of the altar together with the long-necked jars in a location of a *wu*-position (*wu* 午). You must not disrespectfully dump the clay. [97]

Spell to pray for rain (*ch'i-yü chou* 祈 雨 咒): Grand Master of Prime Origin (*T'ai-yüan hao-shih* 太 元 浩 師), Essence of Thunder and Fire, combine *yin* and gather *yang* (yin-yang 陰 陽), keep guard on the Thunder Wall (*lei-ch'eng* 雷 城)[98]. Earl O (*O-po* 閼 伯) [99] with wind and fire, do ascend to the Abyss-Hall (*yüan-t'ing* 淵 庭), let the wind blow and lightning abound, arouse the magic forces out of darkness and let them whirl around [Mount] *T'ai-hua* (太 華). Give commands to lords and retainers as the supreme god-emperors have the decree to act most urgently to put away the *yang* (陽) and let rain come down, which

---

95 This statement could make the impression that the preceding matters were administered by secular official. I think that this is not the case but all of this is a Taoist specialist matter that ritual masters take care of.
96 In the Southeast corner of the altar, see above.
97 Notice, these altars are temporarily set up. It takes some care and precautions to dismantle them after the ritual. Thunder Magic does not need the stage of temples or cloisters. The *wu*-position (*wu* 午) refers to a point on the geomantic compass.
98 See F.C.Reiter: "The Discourse on the Thunders", pp. 218-219, esp. note nr. 90, in: JRAS 14/3.
99 Concerning this name, see TT 1220 *Tao-fa hui-yüan* 133.13b, which shows the Earl to be a great general of the Fire Department of the South. His name, unfortunately, is barely readable.

[must] happen immediately. Send forth urgently the dragons and grasp tightly the lightning that they emerge from darkness and pools. Today I have the honour to speak the spell that must be realised speedily as this is a command by the Jade Emperor (*Yü-ti* 玉帝) for you. Whoever ventures to decline [commits] a crime that does not weigh lightly. Act urgently as this is law and order.

The following Spell of Lord Wood (*mu-lang chou* 木郎咒) says: Heaven of *ch'ien*, (*ch'ien-t'ien* 乾天) let your shining stream. Jade Pool of the East (*yü-ch'ih* 玉池) cast off your fire across a distance of ten thousand miles. In the *K'an*- and *Chen*-palaces (坎震宮) the Lord Wood, **(30a)** *T'ai-i* (太乙), the Hero Three- Mountains (*san-shan hsiung* 三山雄), the *K'un*- (坤) Divinity, the Grand Magic Lord of the *Sun* (巽) Region and the Most Supreme Lady of Purple Vacuity and Mysterious Magic Force, you all must arrive most urgently as this is the law and order.[100]

The spell of the Great Cavern to ask for rain (*Ta-tung ch'ing yü chou* 大洞請雨咒) says: The shining of thunder, the shining of thunder, *T'ai-i* (太乙) is in hiding. *T'ai-po* Director of Wind (*T'ai-po feng-ling* 太伯風令) and Grand Old Lord with Four Eyes (*Ssu-mu hao-weng* 四目浩瓮), Lord Yen (嚴君) you keep guard on the ice and fling away the *yang* (陽) -essences of fire. Director of Rain, for which reason are [your gates] closed? Director of Wind, why do you keep so calm (*p'ing* 平)?[101] The divine might and flamboyance of the sun, gold-immortals awake it. Water and fire flow together in a common stream. Earl Han Po [-wen] (韓伯[文]) bears up his boat, and the three officials (*san-kuan* 三官) open up shining brightness.[102] The five god-emperors (*wu-ti* 五帝) may entrust

---

100   It is not exactly clear what the spell is thought to bring about.
101   I remind of the most likely identity with the names of Feng-po and Yü-shih (風伯雨師) that identify star-divinities, namely Chi-hsing and Pi-hsing (箕星畢星) or numbers 7 and 19 of the 28 stellar divisions, see H.A.Giles: A Chinese-English Dictionary, pp.26,27; and *Sou-shen chi* 4,p.43, ed.: *Ku hsiao-shuo ts'ung-kán*. Chung-hua Comp. Peking 1979.
102   *Fu-chou* (負舟) in this phrase parallels *k'ai-ming* (開明). Possibly this is a reference to Han Po-wen (韓伯文) see TT 1220 *Tao-fa hui-yüan* 133. 12a, which shows the name to be the "radiant and great divinity in the department of lightning". TT 1220: 133 sq present a large and important set of Thunder rituals that have the patronage of T'ai-i (太乙／一). The origin of the phrase "to carry a big boat on the back" most certainly is *Chuang-tzu, nei-p'ien: Hsiao-yao yu* (莊子內篇逍遙遊), see TT 670 *Nan-hua chen-ching* 1.1b., or *Chuang-tzu chi-chieh* 1.1, in: *Chu-tzu chi-ch'eng* 3. See B.Watson: The Complete Works of Chuang Tzu, p.29 "If water is not piled up deep enough, it won't have the strength to bear up a big boat". New York 1968. The sentence features with symbolic expressions the unbearable situation of a drought.

Altars and Prayers for Rain or Clear Skies 105

to T'ai-shang the decree and order that the dragons [must] bring the rain to drip down. Act urgently as this is a decree by the Perfect Lord of Prime Existence (Yüan-ming chen-chün 元命真君).

The spell and prayer for a clear sky (Ch'i-ch'ing chou 祈晴咒) [103] says: Fire Carriage, Fire Carriage (huo-chü 火車), home of Thunder orders (lei-ling 雷令), Three-and-Five (san-wu 三五) are the marshals, shining radiantly without any limit.[104] Let **(30b)** fire speed across a distance of ten thousand miles that can clean away clouds and dawn. You have a bond (yüeh 約) to support the god-emperors with your radiant eyes, silver teeth and the water of wide clarity (huo-luo shui-lei 豁落水類).[105] Be quick and without any delay. The god-emperor [Teng] Po-wen ([鄧] 伯溫)[106] commands to sweep off and annihilate the hidden evil forces (yin-hsieh 陰邪). Act most urgently and speedily and administer immediately the flamboyance of the sun, all like the assistance of the Director of Fire of the supreme god-emperors.

There is an alternative title for this spell. We see it in small print below the main title Spell and Prayer for a Clear Sky. It reads as follows: Spell to put away water and procure clear weather. One has to recite the spell in silent meditation eighty one times.

Subsequently there is still another spell that is used to achieve a clear sky. It says: Essences of metal and jade return,[107] the heavenly clouds may open up and scatter. [The astral constellation] Purple Subtlety (tzu-wei 紫微)[108] may send

---

103 The commentary in small print gives an alternative name: Spell to collect the water and make a clear sky that the priest should silently recite eighty one times. This spell actually addresses Teng Po-wen (鄧伯溫) who is a mighty thunder divinity, see below.
104 Concerning three names or titles of Thunder deities with the specification Three-and-Five, see TT 1220: 56.6b,9b of the pantheon.
105 Compare P.Andersen: "The Practice of Bugang", translating huo-luo豁落 with „wide clarity", in: Cahiers d'Extrême-Asie 5, p.36.
106 See F.C.Reiter: „A Preliminary Study of the Taoist Wang Wen-ch'ing", pp. 174-175. In: ZDMG 152/1. Teng Po-wen (鄧伯溫) is a most prominent name, also in later materials. Compare, for example, TT 1220: 133.13b (T'ai-i chen-lei p'i-li ta-fa 太乙真雷霹靂大法). Here he appears to be the „leading marshal of Thunder and thunderclaps"(Lei-t'ing chu-shuai yen-huo lü-ling ta-shen Teng Po-wen 雷霆主帥焱火律令大神鄧伯溫).
107 The meaning is not quite clear. Possibly this is a reference to sacrificial offerings.
108 See G.Schlegel: Uranographie Chinoise, p.830, nr.699. See H.A.Giles: A Chinese-English Dictionary, nr.12329. See the titles TT 15 Wu-shang chiu-hsiao yü-ch'ing ta-fan tzu-wei hsüan-tu lei-t'ing yü-ching and TT 1485 Tzu-wei tou-shu. Concerning these titles see K.Schipper, p. 1091; and M.Kalinowski, pp.758-759, in: Companion.

down the ominous sign that rain clears off the dust. Earl of the Sea (*hai-po* 海 伯), wind, rain and the radiance of thunder may descend and urge on the essences of *yang* (陽 精) to be most luminous, and the shades of *yin* (陰) may be buried deeply. Urgently and respectfully receive the most exalted and supreme (*t'ai-chi t'ai-shang* 太 極 太 上) fire-wheel-order (*huo-lun lü-ling* 火 輪 律 令) of [the astral constellation] Purple Subtlety (*tzu-wei* 紫 微) in [the heaven of] Jade Purity (*yü-ch'ing* 玉 清).

At this point we find the description of still another amulet: Amulet that Brightens the Eyes with the Character Radiance (*ming-mu kuang-tzu fu* 明 目 光 字 符). [109] The text gives the following instruction: First, you freeze your spiritual forces and pause your breathing. You turn yourself towards the Southeast where the prosperous region is. You hold firmly the writing brush, grind your teeth, grasp and inhale **(31a)** the radiance of the heavenly eye (*t'ien-mu* 天 目) [110] that you expel [from your mouth], pooh, onto the tip of your writing brush. Then it is that you put down the brush to write the amulet.

We get some commentaries in small print concerning the individual elements of the amulet: I respectfully receive the decree from the ruler of the stars, the Great God-Emperor of Purple Subtlety. [111] The *yang* (陽) is the sun, and the *yin* (陰) is the moon. Day and night, they shine brightly. The Jade-Bar (*yü-heng* 玉 衡) [112] revolves and all around there are birth and completion. The left eye flashes with lightning and the right eye has [the shining of] a meteor. The divine forces are pure, and the breaths are lively. They thoroughly penetrate Hades. Sun and moon link up their radiance. Heaven and earth are thus united.

After the text, we find again that fat round dot with a tail, most certainly pointing to the direction of Southeast as we had seen on **p.19b**. Actually, this is the appearance of the amulet after completion – it disappears under the black seal.

---

109   Concerning the practical importance of the title, see pp.31a-31b.
110   A description of the "heavenly eye" gives TT 1220: 69.11b (*Wang Shichen qidao baduan jin* 王 侍 宸 祈 禱 八 段 錦), positioning the heavenly eye between the eyebrows as the magic third eye and focus that emits spiritual might. This refers to the magic deva-eye that the performing priest had obtained by meditation. He adopts, in other words, a heavenly identity when writing out the amulet. Compare *t'ien-yen* (天 眼) as a Buddhist term, see Ting Fu-pao (丁 福 保): *Fo-hsüeh ta-tz'u-tien* 佛 學 大 辭 典), p. 472b. Taipei 1974; see W.E.Soothill and L.Hodous: A Dictionary of Chinese Buddhist Terms, p. 146a.
111   In this case, Purple Subtlety refers to the respective star-divinity.
112   This most likely means the *pei-tou* (北 斗) constellation.

There are some further technical instructions for the elaboration of the amulet. We learn that one has to use for the amulet thin and yellow paper and black ink (*mo* 墨). You do the writing on the morning of the first moon (*shuo-jih* 朔 日) of each month. You burn incense and face the radiance of the sun in the direction of Southeast. You grasp then the breaths and [inhaling] take them in. Secondly, you write the character "light of the sun" (*hsüan* 烜), burn then [the paper] and put [the ashes] into purified **(31b)** water (*ching-shui* 淨 水) [and use it] to wash your eyes. The shining of your eyes will be double as much as usual.

The following method deals with the consecration of incense and its inner application by swallowing (*chou hsiang t'un-fu fa* 咒 香 吞 服 法). The spell consists of the following words: Incense of the law,[113] do enter my body (*ch'ü* 軀) and any illness will be expelled. The law abides, and the incense abides. The breaths abide, and the spiritual agents with their might abide. When they change (*pien* 變), they thus become white frost. When they transform (*hua* 化), they thus become sweet dew. The incense of the law may always be my ruling force for 36000 days. Act urgently as this is the law and order of T'ai-shang Lao-chün (太 上 老 君).

At this point, the text gives us an instruction for the application of the spell. Each morning we have to take one portion of frankincense (*ju-hsiang* 乳 香) and inhale grasping the radiant breath of the sun. Three times, we exhale the breath [loudly] coughing above the incense and use purified water to swallow it.[114] We can preserve our body to be fine and without any illness for one full year, and our eyes will have the penetrating eyesight.

The next theme is a prayer for snow (*ch'i-hsüeh* 祈 雪): **(32a)** We read the following statement by the Fire Master (*huo-shih* 火 師): Generally, the spring is warm in the world of man, the summer is hot, the autumn is cool and the winter is cold. Now, this is the constant principle of *yin* and *yang* (陰 陽) that rise and descend. When the four seasons are in harmony, evil breaths (*hsieh-ch'i* 邪 氣) do not come to life (*pu-sheng* 不 生), and illnesses and epidemics do not arise. When one season is not quite in tune (*pu-ying* 不 應), the people suffer from disasters as a response. When the winter is warm and there is no snow at all, this is called "*yang* battles" (*yang-chan* 陽 戰), which causes people to be variously ill during springtime. When governors (*kuo-wang* 國 王) and common locals (*t'u-jen*

---

113 Alternatively translate: „incense of the ritual". The translation derives from my interpretation of the last sentence of the spell.
114 This may refer to the ashes of the incense.

土人) ask for snow [to fall] and for the expulsion of disaster, they report then the situation by means of *tz'u* (詞) [-prayers]. They are sent off as soaring memorials to the supreme god-emperors to notify the Court of the Emissaries of the Five Thunders and the Three Monitoring Offices of Thunder and Thunderclaps (*Lei-t'ing san-ssu* 雷霆三司). They also notify the Divine and Supreme Councillor of the East and the Immortal Master of Portentous Radiance (*Tung-ling shang-hsiang Jui-kuang hsien-shih* 東靈上相瑞光仙師). [115] The memorial begs for pardon concerning the transgressions of the living beings (*sheng-ling* 生靈) and implores that portentous snow may be sent down earlier as to suppress the evil breaths. Then you go to a location that corresponds above [in heaven] with the *T'ai-sui* [star] (太歲), and exactly in that location you set up an altar with three levels in accordance with the model of the altar that is used to pray for rain. You order the Taoist priests (*tao-shih* 道士) to do everything just in the same way as for the rituals [described] before. It is, however, required that in the centre of the altar and above the incense table for the ritual veneration you paint two icons (*liang-wei* 兩位) in order to venerate the Divine and Supreme Councillor of the East and the Immortal Master of Portentous Radiance. [116] **(32b)** Every time the ritual master comes to the three points of time, namely *ch'en, wu* and *yu* (辰午酉) he performs the [ritual] steps "produce water and summon thunders" (*tso-shui chao-lei kang* 作水召雷罡). He holds firmly his sword and ascends the altar. Turning to the right hand side, he walks around [the altar] thirty six times and recites silently (*mo-nien* 默念) thirty six times the Spell of the Eastern Divine Force (*tung-ling chou* 東靈咒). When he reaches the front side of the incense altar, he presents incense and makes his statement to the supreme god-emperors, to the Court of the Emissaries of the Five Thunders, to the Three Monitoring Offices of Thunder and Thunderclaps (*lei-t'ing san-ssu* 雷霆三司), to the Divine and Supreme Councillor of the East and the Immortal Master of Portentous Radiance (*Tung-ling shang-hsiang Jui-kuang hsien-shih* 東靈上相／瑞光仙師), [117] that they may soon send down portentous snow.

---

115 These are two addressees. Compare TT 15 *Wu-shang chiu-hsiao yü-ch'ing ta-fan tzu-wei hsüan-tu lei-t'ing yü.ching* 17a, which at first sight suggests that this is just one name and title. However, see below the reference to the two icons that have to be set up. They stand for two identities. The pantheon in TT 1220: 56.5b has Tung-ling shang-shang yüan-chün (東靈上相元君) and p.6a has Jui-kuang hsien-shih (瑞光仙師).

116 A commentary in small print points out that the name *Tung-ling* (東靈) in some other text reads "the Eight Magic Forces of the Eastern Pole" (*dong-chi pa-ling* 東極八靈).

117 Concerning the two names, see above.

Now, following this and relying on the formal procedures you can descend from the altar. It will not take more than the time of one prostration and there will be a response. The ritual to thank for the snow follows the formal arrangements of the ritual for prayers for rain.

The Spell of the Eight Magic Forces of the Eastern Pole (*tung-chi pa-ling chou* 東極八靈咒) says: Divine Supreme Councillor of the East, home of portentous radiance, you secretly keep in control the heavenly flowers of green jasper of the superior god-emperors. You rule over the wide and clear fire and let the wind sweep severely in order to eliminate the evil [forces]. Your compassion and benevolence save the creatures, responding in time without failure. I desire that you send down portentous snow that may spread all over China (*chung-hua* 中華). Act urgently as this is law and order. **(33a)**

Another spell says: Gold-Essences, Great Ultimate, Jade Flowers that disperse their leaves, I [modestly] pray for snow and pray so to move away the warm atmosphere of wind and fire. Luminous Master, Fire Master (*huo-shih* 火師), do let portentous [signs] appear. Superior Perfected Ones in the Great Void, in [the Heaven of] Great Purity and at the Ultimate Poles (*t'ai-hsü t'ai-ch'ing t'ai-chi* 太虛太清太極), do assist [the execution of] this order.

The Crusade against Temples and the Eviction of Wicked Spirits (*fa-miao ch'u-ching* 伐廟除精)

The Fire Master says: You respond to life among people when there are outrageous spirits, fierce demons, mountain spirits and monstrous demons (魅) that lead astray common wives and daughters, let their bodily appearances become visible and secretly rob properties of the people. Thus, they have shrines and sow the seeds to commit transgressions. Thunder and thunderclaps supervise the investigation of such elements, because they do not fear the laws of the immortals (*hsien-fa* 仙法). It is well permissible that people expose *tz'u* [-prayers] (*t'ou-tz'u* 投詞) to memorialize all the harm [they have suffered]. They rely then on these memorials to inform the superior god-emperors and send off soaring notifications to all the officials, using special documents for the Divine Thunder and the Water Thunder (*shen-lei shui-lei* 神雷水雷), which may convene and proceed [together], crusading against [such evil elements] within a certain time.

When the right time has come, you carry on your left forearm the Seal (*yin* 印) of Thunder Radiance and Fire Script (*lei-kuang huo-wen* 雷 光 火 文) and the Seal **(33b)** of Mercury Heaven of Purple Radiance (*tzu-kuang tan-t'ien* 紫 光 丹 天). You perform the [ritual] steps "the soaring dragon beheads the monsters" (*fei-lung chan-kuai kang*飛 龍 斬 怪 罡). You do so to move the violent and evil forces (*k'uang-hsieh* 狂 邪) and crusade against [the respective] temple gate. You attach first one Amulet of the Great Divinity of Blazing Fire (*Yen-huo ta-shen fu* 焱 火 大 神 符) at the gate. Having done so, you return to some high and exposed location on the right and left sides [of the temple in consideration]. You sacrifice in the open (*yeh* 野) to the thunder divinities (*lei-shen* 雷 神), shout out for them to come forth and approach speedily that site in order to burn down and destroy the temple and to put away and arrest the demonic thieves. Responding right in time there will be extensive thunders and rainfall. Do not be afraid at all. When the ritual procedures are complete, you offer a feast (*chi-ch'ou* 祭 酬) for the Thunder and thunderclaps. When you use this ritual for your performance, you must perform it publicly and correctly. You must not use this method for any erratic endeavours. You have to fear that you hurt creatures, but [in this case] the ritual officer (*fa-kuan* 法 官) invokes all the blame to rest upon him.

The beheading of scaly dragons, sea-serpents and water monsters (*chan chiao-ch'en shui-kuai* 斬 蛟 蜃 水 怪) is the following theme. When water-serpents, dragons and water monsters dwell in rivers, lakes, ponds and in dark caves with pools and fountains, when they devour the six [sorts of] domestic animals, suck the blood veins of men and eventually change to be huge serpents that bar the main roads in the area and harm the merchants, or when they attack untimely and cause great floods that inundate provinces and districts, you memorialise all of this precisely **(34a)** vis-à-vis the supreme god-emperors. You notify all officials and send documents to the Water Department (*shui-fu* 水 府), asking for generals and emissaries to be sent down in order to crusade against [these plagues] within a certain time. Three days in advance, you must present warrants (*tieh* 牒) to the city god (*Ch'eng-huang* 城 隍) in the [respective] province or district (*chou-hsien* 州 縣), to the earth-god (*t'u-ti* 土 地), to the village altars (*li-she* 里 社) and the local shrines (*miao-tz'u* 廟 祠), that they group and lead on their [spirit] troops and generals to control the roads in the four directions and prevent the dragons, monsters and evil water spirits to pass through. When the specific time and day have come, you use flags that are made of dark red silk, seven *ch'ih* long, and write out seven pieces of the Amulet Heavenly Barrier (*t'ien-kuan fu* 天 關 符). You use the Seal of Cave Magic of the Jade Morning (*yü-ch'en tung-ling* 玉 晨 洞 靈) to seal the amulets. Carrying the seal on the left upper arm, you form twisting [the fingers] the *mudrâ* Thunder Office (*lei-chü* 雷 局). Your right hand grasps the sword and you perform the [ritual] steps "connect the heavenly iron barricade" (*lien t'ien t'ieh-chang kang* 連 天 鐵 障 罡). You approach the place where the sea serpents and dragons are in hiding and throw the amulets into the water. When you go to expose the amulets [in the water] you must instruct the [local] people to beat brass gongs and drums as to support the action. When the exposure of the amulets had been done, you must summon and invite with a very loud voice all the officials, generals and troops and the legions of the department of the Five Thunders that they speedily behead the water monsters. When [the water monsters] shrink back into the water, the thunders shake the waves that greatly boil up and make these monsters come up to drift on the surface of the water. You most urgently must **(34b)** take out then the Seal Cave-Magic (*tung-ling* 洞 靈) [118] and let it shine on the monsters. [119] The blood of the monsters shall flow out of their two eyes and thus they perish. After this has been achieved you order your attendants (*ti-tzu* 弟 子) to stab [the monsters] with their swords and bury the cadavers in a remote place that must be secured (*chen* 鎮) with [the help of] amulets. When this procedure is complete, you issue amulets in order to send back all the troops and generals to return to their original spheres of might [where they are stationed]. Later then and quite independently, you choose an auspicious day to present sacrificial offerings as thanksgiving.

A Corpus of Taoist Ritual now introduces seven heavenly characters that constitute the dispersed form of the Amulet of the Seven Killers of the Heavenly Barrier (*t'ien-kuan ch'i-sha fu* 天 關 七 煞 符). The description says: **(35a)**: The seven amulets are extremely magic (*chi-ling* 極 靈).[120] They dominate the water monsters of a whole region. In case that you want to bring peace and control to one region or one spot, you write the seal characters of the amulet in a compound structure (*tieh-chuang* 疊 篆). You use iron ore (*sheng-t'ieh* 生 鐵) to cast [the amulet]. When it is completed you place [the amulet] safely at pools and caves and for all future ages there will not be again any harm caused by any water monsters. [The amulet characters] present the decree to behead and eliminate poisonous dragons and water monsters "urgently as this is the law and order".

---

118 This most likely is the above mentioned seal Cave Magic of the Jade Morning (*yü-ch'en tung-ling* 玉 晨 洞 靈)
119 This instructions seems to have a "historical" precedent, involving the two famous Taoists Hsü Sun (許 遜) and Wu Meng (吳 猛), compare above TT 1220: 56.10a- 10b.
120 Notice that each of the seven characters can count as one amulet.

The following subtitle reads: Multilayer seal character that assembles the outer form of the seven killers of the heavenly barrier (*t'ien-kuan ch'i-sha chü-hsing tieh-chuan* 天 關 七 煞 聚 形 疊 篆). Below the seal character of the compound seal, we get the indication that we should add the religious rank, the surname and name [of the performing priest].

Ritual Steps and Mudrâs (*pu-kang chüeh fa* 步 罡 訣 法) [121]

Concerning the theme there are quite a few subtitles: **(35b)** The Fire Master (*huo-shih* 火 師) says that the men who cultivate perfection and receive the ritual methods have to encounter an enlightened master. He would instruct them concerning the pure amulets (*fu-t'u* 符 圖) and the patterns of the precious seals (*pao-yin wen* 寶 印 文) of the Five Departments of Thunder Rituals (*lei-fa wu-pu* 雷 法 五 部). Furthermore, the persons need to get the choreography for the ritual paces and the [proper] measures for the *mudrâs* (*kang-pu chüeh-chieh* 罡 步 訣 節). Only then, it is that the men can attain the magic force [of amulets and seals]. It is for this reason that in correct rituals there are the [spirit] generals *Chien-kang* (建 罡 將 軍) and *Ch'i-kang* (起 罡 將 軍).[122] All [correct rituals] document this matter. Today I transmit it for the posterity.

Now we see five titles and drawings of ritual choreographies that show the choreography of various ritual paces.

Ritual paces to produce water and summon the thunders (*tso-shui chao-lei kang* 作 水 召 雷 罡).[123]

Ritual paces of the soaring dragon that brings rain (*fei-lung chih-yü kang* 飛 龍 致 雨 罡). **(36a)**

Ritual paces of the soaring dragon that beheads monsters (*fei-lung chan-kuai kang* 飛 龍 斬 怪 罡).

---

121 *chüeh* (訣) most certainly means "hand-gesture" (*mudrâ*), see below pp. 36b-37a.
122 The two spirit generals most likely represent the stars at the tail and the head of the Big Dipper. They point to the crucial orientation of proper Thunder rituals.
123 *kang* (罡) refers to the astral constellation of the Big Dipper that provides the standard orientation for the choreography of the ritual steps. The translation "ritual paces (or steps)" implies the astral orientation.

Ritual paces that drive on the dragons to suck up the flood (*ch'ü-lung hsi-shui kang* 驅龍吸水罡) **(36b)**.

Ritual paces of the iron barricade that connects heaven (*lien t'ien t'ieh-chang kang* 連天鐵障罡).

All the drawings give a rather detailed mapping of the paces. The first drawing most clearly shows that the choreography follows the basic pattern of the Big Dipper.

The text continues to present a conclusive list of hand-gestures (*mudrâs/chüeh-mu* 訣目) that contains the following items:

The *mudrâ k'uei-mu* 魁目 is the *hsü* (戌) -pattern [124] that must be pressed with the thumb of the left hand.

The *mudrâ kang-mu* 罡目 is the *ch'en* (辰) -pattern.

The *mudrâ shen-mu* 神目 is the *wu* (午) -pattern.

The *mudrâ kuei-mu* 鬼目 is the *tui* -pattern with *mao* on top of it (兌卯).

The *mudrâ* "barring the wicked influences of the wind" (*chin feng-hsieh* 禁風邪) requires that the thumb of the left hand presses upon the second finger.

The *mudrâ* "seal of comprehensive assistance" (*tsung-she yin* 總攝印) requires that [the thumbs of] both hands press upon the middle section of the [respective] middle fingers.

The *mudrâ* "seal for the employment of emissaries" (*i-shih yin* 役使印) means that a position between the positions *mao* and *ch'en* (卯辰) [must be pressed upon].

(37a) The *mudrâ* Thunder Office (*lei-chü* 雷局) means that the thumb presses the second and the third fingers and forms twisting the *tzu* (子) -pattern. [The same time] the fourth and fifth fingers press upon the centre of the palm (*chang-hsin* 掌心).

---

124 Alternatively "line" that has to be pressed upon, see below. Pattern translates the word *wen* (文).

The following Text on Transforming to be a Divine Entity (*hua-shen wen* 化神文) certainly makes some statements that have a top priority for Taoist Thunder rituals. The commentary in small print reveals the importance of the title. It points out that generally any writing of amulets, any dispatch of literary documents and any practice of self-cultivation (*hsing-ch'ih* 行持) [125] require to transform the own person to be a divine entity.

The Fire Master (*huo-shih* 火師) says that any gentleman who performs Thunder rituals (*lei-fa* 雷法) and, at any given time, has the task to expel [wicked elements], employ [divine forces], address and call out [for divine help], would have to use the *mudrâ* "transform to be a divine entity" (*pien-shen chüeh* 變神訣). He grinds his teeth five times (*k'ou-ch'ih* 叩齒) and visualises (*ts'un* 存) that he wears on the top of his head a *liang*-cap (*liang-kuan* 梁冠), is clad in red garment and wears red shoes. [He visualizes] immortals who on his left and right sides hold up streamers. There are [also] young lads who present respectfully the sword and keep the registers (*lu* 籙) at hand. Other young lads hold up banners. There are the judges, who keep the registers of evil deeds. There are the generals of the Thunder lords (*Lei-kung* 雷公) of the Five god-emperors (*wu-ti* 五帝), the Thunder Lord *Shao-yang* (*Shao-yang lei-kung* 卲陽雷公), the Generalissimo of the Fire Carriages (*huo-chü yüan-shuai* 火車元帥) and the divine generals of all the [Thunder] offices. They altogether appear in front of me, [126] behind me and on my left and right sides. They listen to my orders and directives. After this, I start to perform the ritual [in consideration].

The following text presents a Spell for Dispatching Texts and Scripts (*fa-ch'ien wen-tzu chou* 發遣文字咒) saying: I respectfully receive the command from the god-emperors and keep seals and amulets in tight control. Having the sole power of command over the Headquarters Offices [127] I command the affairs of demons **(37b)** and divinities who do not operate in the right order. The execution of my order must be as fast as the fire of the stars (*hsing-huo* 星火).

---

125 This is the title of the third paragraph in *Wang Shih-ch'en ch'i-tao pa-tuan chin* (王侍宸祈禱八段錦) in TT 1220: 69.11a-14a, where the term clearly points to the type of self-cultivation and meditation that has to precede any ritual action in the framework of Thunder Magic.

126 Explicitly *wu* (吾), representing the Thunder specialist. For the two divinities Thunder Lord *Shao-yang* (*Shao-yang lei-kung* 卲陽雷公), Generalissimo of the Fire Carriages (*Huo-chü yüan-shuai* 火車元帥), see the pantheon in *chapter 56*, TT 1220: 56.9a-9b, where also the attendant lads are mentioned. P.6b gives a similar title, adding Three-and-Five.

127 See *Hucker*, p.542, nr.7285.

When the ritual [force] (*fa* 法) spreads out, it urgently leads wind and thunderclaps to victory. The three spheres (*san-chieh* 三 界) must not let the slightest delay happen, and within shortest time there are bright responses. Obey and accept the respective commands and do not hastily delay or retain them (*ch'ih-liu* 遲 留). Go quickly and return quickly. I attend.

Another spell says: Heavenly thunders, all in hiding, dragon and tiger interlock (*lung-hu chiao-heng* 龍 虎 交 橫). Sun and moon spread out your radiance. Shine on me and give me a share of your brightness. Palace Guard Emissaries, attend to the amulet, accept and execute [my orders].

The subsequent text presents an instruction concerning sacrifices that are performed in open air (*yeh-chi shen-chüeh* 野 祭 神 訣).[128] The Fire Master (*huo-shih* 火 師) gives the following instruction: Generally, when prayers and requests are made in order to crusade against a temple (*fa-miao* 伐廟) to expel wicked and cruel [elements], and when a fast response is desired to come about it is necessary to use the method of sacrificing in open air. In just a moment, there will be an appropriate response by the lower generals of the Thunder department (*lei-pu* 雷 部), who are wild and fierce. You have to offer them blood sacrifices (*hsüeh-shih* 血 食). The suitable ritual sacrifice has to be staged in open air. You choose a spot on a high mountain (*kao-shan* 高 山) **(38a)** that nobody else reaches. [If the ritual is performed] at a market place (*shih* 市) you must use a single [separate] room in the position of Southeast, lock the room and present the sacrifice [inside]. When you set up the altar you take white chalk and mark out the altar [on the floor] with its three terraces facing [the direction of] sun (巽). The upper platform [of the altar] is broad nine *ch'ih*, the one in the middle is broad one *chang* and two *ch'ih*, and the lower platform is broad one *chang* and five *ch'ih*. You take one cock and fix firmly the legs with a dark red string of silk. You take five big wine-cups and one bottle of wine, five thousand [pieces of] sacrificial money and five sacrificial flags, one sharp-edged sword and frankincense for one incense burner. You arrange the altar according to the rules.

When the time has come, the ritual master (*fa-shih* 法 師) wearing his hair dishevelled and barefooted looks far out into the direction of Southeast and performs the ritual steps "crush the earth and summon the magic forces" (*p'o-ti chao-ling kang* 破 地 召 靈 罡). He ascends the altar and sacrifices incense (*shang-*

---

128 Most certainly, this does not just mean outdoors but points to a remote spot in the wilderness.

hsiang 上 香). His left hand forms twisting the *mudrâ* Employ-the-Emissaries (*i-shih chüeh* 役 使 訣), and his right hand holds the sword. Again, looking out into the direction of Southeast the ritual master declares with a loud voice: I respectfully receive the decree by the god-emperors to summon the Thunder Emissaries of Barbarian Thunder of the Five Directions (*wu-fang man-lei shih-che* 五 方 蠻 雷 使 者), that they approach speedily to so-and-so location, to punish and behead the demonic thieves [at the location].

Perhaps the ritual master also says: Heaven, send down sweet rain and order wind and fire to be around quickly. Today I have prepared, heeding the rules, the blood of a cock, wine, sacrificial money, bright incense, sacrificial presents and other items. The time has come and nobody must offend against **(38b)** my orders. When the spell has ended, you stab the cock and let the blood drip into the wine-cups that stand for the five directions. You pour then wine into the five cups and use your sword to stir and mix the wine [with the blood] in the wine-cubs. It is all the same as with the preceding spell. You burn (*hua* 化) the sacrificial money and take the sacrificial offerings to bury them three *ch'ih* [deep] in the earth [at a spot] in the direction of Southeast (*sun* 巽). The ritual master (*fa-shih* 法 師) speaks again a spell saying: I am in front of the altar and my eminent sword is at avail. He takes his sword and sticks it upright into the earth where he had buried the sacrificial offerings, and then he retreats. Clouds will immediately come up and rain will start to fall. You must not often sacrifice [in this way] and hurt creatures. At the time of the sacrifice, you had one amulet fastened at the centre of the altar (*t'an-hsin* 壇 心). When the ritual has come to its end, the amulet must be buried together with the [other] sacrificial objects. This method is extremely subtle and must not be leaked [to outsiders] because one has to be afraid that disaster may be caused to happen.

Now, we find a drawing that shows the altar with the flags as described in the text. There are two lines of Chinese characters in small print running down from the top to the bottom line on both sides of the square drawing. The lines indicate where the surname and name of the respective Emissary of the Barbarian Thunder can be inserted. The columns end with a statement saying, for example, "white garment and cap" (*i-tse* 衣 幘), which speaks about the emissary of the West. The phrases black garment and cap, **(39a)** red garment and cap, green garment and cap point to the other three directions.

Conditions for the Ritual Transmission and [spirit] Promotion (*ch'uan-tu t'iao-p'in* 傳度條品)

Here we read: generally, the salvation of several hundreds of people counts as one merit completed (*i-kung* 一 功), also to pray for rain at one time counts as one merit completed. The achievement of [these] highest merits leads exactly to one promotion (*chuan* 轉).

At the start, the respective teacher master transfers the *Shang-ch'ing* Register that Preserves Clear Measures (*pao ming-tui shang-ch'ing lu* 保 明 兌 上 清 錄)[129] to the disciple of Thunder Magic (*lei-fa ti-tzu* 雷 法 弟 子), to let him serve as Judge on the Right Side at the Five Thunders Court (*wu-lei yüan yu p'an-kuan* 五 雷 院 右 判 官). When the merit [on this position] is complete, the person[130] is promoted to [the rank of] Judge on the Left Side (*tso p'an-kuan* 左 判 官). Then the person is promoted to [the rank of] Great Judge on the Right Side (*yu ta p'an-kuan* 右 大 判 官)[131] and the same time Chief Corrector of the Proper Rituals and Lord of Origin for the [respective] day (*cheng-fa chu-ko jih-chih yüan-chün* 正 法 主 格 日 直 元 君).

The following promotion is [the rank of] Great Judge on the Left Side (*tso ta p'an-kuan* 左 大 判 官). When the merits [on the position] suffice, [the respective person] is promoted to [the rank of] Commissioner of the Capital Waters at the Dipper (*tou-chung tu-shui shih-che* 斗 中 都 水 使 者). The commentary below in small print names still another title namely Commissioner that Controls the Waters (*chang-shui shih-che* 掌 水 使 者). Furthermore, the person is promoted to [the rank of] Commissioner of the Capital Waters (*tu-shui shih-che* 都 水 使 者).[132]

Another commentary in small print indicates for this level the two alternative titles namely Judge on the Right Side for the Capital Waters and secondly Judge on the Left Side for the Capital Waters (*tu-shui yu/tso p'an-kuan* 都 水 右 / 左 判 官). When the merit on that position suffices, again the rank rises to be **(39b)**

---

129 Concerning *tui* 兌, see H.A.Giles: A Chinese-English Dictionary, nr.12170. It is very unlikely that the character is used instead of nr.13776 "to be pleased" (悅). *Lu* (錄) certainly stands for "register" *lu* (籙).
130 This is the disciple of Thunder Magic, who rises to some remarkable spirit ranks.
131 Concerning the titles of these judges, see TT 1220: 56.6b of the pantheon.
132 As to these titles, compare TT 1220: 56.6b of the pantheon.

Arbiter of Fate in the Heaven of Supreme Purity [and] Councillor on the Right Side in the Jade Department (*shang-ch'ing ssu-ming yü-fu yu-ch'ing* 上清司命玉府右卿). The commentary in small print says that the title is identical with the Imperial Secretary in the Department of Fire (*huo-pu shang-shu* 火部尚書).

Again, the rank changes to be Arbiter of Fate in the Heaven of Supreme Purity, Councillor on the Left Side in the Jade Department (*shang-ch'ing ssu-ming yü-fu tso-ch'ing* 上清司命玉府左卿) that again the commentary says to be identical with the Imperial Secretary in the Department of Fire (*huo-pu shang-shu* 火部尚書).

The following promotions are the ranks: Superior Councillor at the Jade Department, Associated Emissary and Judge of the Five Thunders (*yü-fu shang-ch'ing wu-lei fu-shih p'an* 玉府上卿五雷副使判),[133] Supervisor of Public Affairs of Demons and Deities at the Headquarters Office of Thunder and Thunderclaps (*lei-t'ing tu-ssu kuei-shen kung-shih* 雷霆都司鬼神公事).

The next promotion confers the title Superior Councillor at the Jade Department, Emissary of the Five Thunders, Concurrent Supervisor of Public Affairs of Demons and Deities at the Headquarters Office of Thunder and Thunderclaps (*yü-fu shang-ch'ing wu-lei shih ling lei-t'ing tu-ssu kuei-shen kung-shih* 玉府上卿五雷使領雷霆都司鬼神公事). The commentary in small print says that the title is identical with the title [and function of] Censor at the Golden Palace in the Nine Heavens (*chiu-t'ien chin-ch'üeh yü-shih* 九天金闕御史) that also corresponds with the controlling function of the Judge on the Right Side and the Judge on the Left Side. The position may be called Joint Administrator and Supervisor, Manager of the Public Affairs of Demons and Deities at the Headquarters Office of Thunder and Thunderclaps (*t'ung kuan-kan lei-t'ing tu-ssu kuei-shen kung-shih* 同管幹雷霆都司鬼神公事). The official at the Jade Department (*yü-fu ch'ing* 玉府卿) may be called Supervisor and Administrator of the Public Affairs of Demons and Deities at the Headquarters Office of Thunder and Thunderclaps (*chih lei-t'ing tu-ssu kuei-shen kung-shih* 知雷霆都司鬼神公事).[134]

The title Tablet of Declaration of Thunder and Thunderclaps (*lei-t'ing pan-kao* 雷霆版誥) introduces the drawing of an amulet that has to be done in red ink on a plank made of the wood of pine trees (*tzu-po* 梓柏). We first see the front

---

133   For this title, compare TT 1220: 56.5a of the pantheon.
134   We notice that these divine titles do not matter in the preceding texts.

side of the plank that must be broad three *ts'un* and long five *ts'un*. We get the formulae for the two inscriptions on the front side and on the backside of the wooden plank. A text in small print below the drawing of the amulet says that so-and-so disciple can be entrusted with this or that task, which he will perform in accordance with the standards.

(**40a**) The backside of the plank shows in the centre line the date of issue of the declaration. The right line names the [acting spirit] emissary who has so-and-so name and holds the rank of Emissary at the Court of the Jade Pivot (*yü-shu yüan shih* 玉 樞 院 使).[135] The line on the left side of the central line with the date of issue indicates the street, the rank and the names of the respective teacher master.

A similar presentation of the Iron Bond of the Thunder Wall (*lei-ch'eng t'ieh-ch'üan* 雷 城 鐵 券) follows. The central line of the text on the front side says: Decree, do guard the body and keep safe the life. Employ and dispatch the Five Thunders. This has to be done in red ink. On the right side, we get eight characters of *dhâranî* spells that control the [spirit] palace guards of all departments who record the merits and the weight of transgressions. On the left side of the central line, we read two *dhâranî* characters. The subsequent text says: Urgently! All divine forces may accept and act according to [the amulet]. As far as the teacher master is concerned, any crime report about him being remiss in ritual matters reaches the three officials (*san-kuan* 三 官) for his impeachment. - The next line to the left belongs to the backside of this bond and again names the Emissary at the Court of the Jade Pivot (*yü-shu yüan shih* 玉 樞 院 使), [who in fact is] so-and-so official with so-and-so surname. The formulae on the backside are identical with the respective text of the Tablet of Declaration of Thunder and Thunderclaps that we have just seen.

These rather formalistic presentations are very much in line with the preceding descriptions of spirit ranks that the successful Thunder specialist can hold. The administrative instructions round up and conclude *chapter 56*.

I only summarize briefly the two formal descriptions of memorials and administrative documents that follow (**pp.40a-42a**). The standard formulae enlist the help of the Court of the Emissaries of the Five Thunders in the Jade Department of the Heaven of Highest Purity (*Shang-ch'ing yü-fu wu-lei shih-yüan* 上 清 玉 府 五 雷 使 院). All the available [spirit] military forces, the amulets,

---

135 This is the spirit position that the priest has adopted to work out the amulet.

seals, *mudrâs* and emissaries are to forward the report or petition (*shen* 申) of the administering priest, who calls for the support of the highest spirit-authorities in heaven to help the country and save the people. Any support must be provided due to the orders of the supreme god-emperors, who receive the petition by the intermediaries of the Jade Department. There must be no offence against or suppression of the requests brought to the knowledge of the authorities at the Jade Department by means of the basic prayer tablets that the sincere followers of Taoism exposed. Individual protection and salvation is also asked for, and this extends to the salvation of the seven generations of ancestors in Hades, which again is a standard requirement in terms of filial piety.

The first document has the specific title: Form for Memorials (*cha-tzu shih* 劄子式). (**40b**) The second sketchy document gives the basic pattern for an address to the Court of the Heavenly Pivot (*shen t'ien-shu yüan* 申天樞院). First, there is an appeal to the subordinate Court of the Emissaries of the Five Thunders in the Jade Department of the Heaven of Highest Purity (*Shang-ch'ing yü-fu wu-lei shih-yüan* 上清玉府五雷使院). In each case, we learn how to date, sign or seal the documents. ***

Abbreviations

AAS: Asien- und Afrika-Studien der Humboldt Universität zu Berlin. Harrassowitz/Wiesbaden

AF: Asiatische Forschungen. Harrassowitz/Wiesbaden

Book of Changes: The I Ching or Book of Changes, the Richard Wilhelm Translation rendered into English, by C.F.Baynes. New York 1967.

Companion:: K.Schipper and F.Verellen eds.: The Taoist Canon, A Historical Companion to the *Daozang*. 3 vols.Chicago 2004.

Hucker: A Dictionary of Official Titles in Imperial China. Stanford 1985.

JRAS: Journal of the Royal Asiatic Society

MOS: Münchener Ostasiatische Studien. Steiner/Stuttgart

TT: Taoistischer Kanon (*Ta Ming Tao-tsang ching/Cheng-t'ung tao-tsang*). Ed. Taipei 1977, 61 vols.

ZDMG: Zeitschrift der Deutschen Morgenländischen Gesellschaft

Bibliography

P.Andersen: "The Practice of Bugang", in: *Cahiers d'Extrême-Asie* 5, pp.15-53 (1989-1990).
P.Andersen and F.C.Reiter eds.: Scriptures, Schools and Forms of Practice in Daoism, A Berlin Symposium, in: AAS 20 (2005).

J.M.Boltz: A Survey of Taoist Literature, Tenth to Seventeenth Centuries. Berkeley 1987.

Chang Chih-hsiung/Li Feng-mao: *Cheng-yi fa-fu yü tao-chiao wen-hua. Shou-chieh hai-hsia liang-an tao-chiao wen-hua lun-t'an.* Chiang-hsi 2005.
Kwang-Chih Chang: "China on the Eve of the Historical Period", pp.37-73. In: M.Loewe and E.L. Shaughnessy eds.: The Cambridge History of Ancient China, From the Origins of Civilization to 221 B.C. Cambridge 1999.
Ch'en Kuo-fu: *Tao-tsang yüan-liu k'ao.* Rpr. Taipei 1975.
Ch'en Shun-yü: *Lu-shan chi*, in: *Pi-chi hsü-pien*. Ed. Taipei 1969.
Chou-i, ed.: *Kanbun taikei*. Tokyo 1913.
Chuang-tzu chi-chieh, in: *Chu-tzu chi-ch'eng. Peking 1986.*

W.Eichhorn: Die Religionen Chinas. Stuttgart 1973.

S.D.R.Feuchtwang: An Anthropological Analysis of Chinese Geomancy. Vientiane 1974.

H.A.Giles: A Chinese-English Dictionary. Rpr.Taipei 1972.

Hsiao T'ien-shih : *Hsü-hsien chuan*, in: *Li-tai chen-hsien shih-chuan, Tao-tsang ching-hua* 5/7. Taipei 1980.
Hsü Chien: *Ch'u-hsüeh chi*. Ed. Peking 1980.
Huang K'an ed.: *Pai-wen shih-san ching, Shang-shu; Chou-shu hung-fan.* Shanghai 1983.
R.Hymes: Way and Byway. Taoism, Local Religion and Models of Divinity in Sung and Modern China. Berkeley 2002.

Kan Pao: *Sou-shen chi*. Ed. *Ku hsiao-shuo ts'ung-k'an*. Chung-hua Comp. Peking 1978.
L.Kohn ed.: Daoism Handbook. Leiden 2000.

J.Lagerwey: *Wu-shang pi-yao*, somme taoiste du VI siècle. Paris 1981.
-: Taoist Ritual in Chinese Society and History. New York 1987.

M.E.Lewis: "Warring States Political History", p.632, in: M.Loewe E.L.Shaughnessy eds.: The Cambridge History of Ancient China, From the Origins of Civilisation to 221 B.C. Cambridge 1999.
Li Fang : *T'ai-p'ing guang-chi*. Ed. Kyoto 1972.
Li Yüan-kuo: *Shen-hsiao lei-fa*. Chengdu 2003.
—: "*Tao-chiao lei-fa yen-ko k'ao*", in: *Shih-chieh tsung-chiao yen-chiu*. 2002, 3, pp.88-96.
—: "*Lun tao-fu te chieh-kou yü pi-fa*", in: *Tsung-chiao hsüeh yen-chiu*. 1998, 2, pp.8-13.
Liu Chung-yü: *Tao-chiao fa-shu*. Shanghai 2002
M.Loewe and E.L. Shaughnessy eds.: The Cambridge History of Ancient China, From the Origins of Civilization to 221 B.C. Cambridge 1999.
P. van der Loon: "A Taoist Collection of the Fourteenth Century", in MOS 25: W.Bauer ed.: Studia Sino-Mongolica, pp.401-405 (1979).

H.Maspero: Le taoisme et les religions chinoises, p.535 (Les procédés de nourrir le principe vital). Paris. Rpr. 1971.
R.Mathieu: Étude sur la mythologie et l'ethnologie de la Chine ancienne, traduction annotée du Shanhai jing. Paris 1983.

P.Nickerson: "Attacking the Fortress, Prolegomenon to the Study of Ritual Efficacy in Vernacular Daoism", pp.117-179, in: AAS 20 (2005).

P'ei-wen yün-fu, ed. Taiwan Shang-wu Comp. Taipei 1966.

F.C.Reiter: The Aspirations and Standards of Taoist Priests in the Early T'ang Period, in: AAS 1 (1998).
—: The Management of Nature: Convictions and Means in Daoist Thunder Magic (Daojiao leifa), pp.193-210, in: AAS 29 (2007).
—: "The Name of the Nameless and Thunder Magic", pp.97-116, in: AAS 20 (2005).
—: Religionen in China, Geschichte, Alltag, Kultur. München 2002.
—: "The Discourse on the Thunders, by the Taoist Wang Wen-ch'ing (1093-1153)", in: JRAS 14/3, pp.207-229.
—: "A Preliminary Study of the Taoist Wang Wen-ch'ing (1093-1153) and his Thunder Magic (lei-fa)", in: ZDMG 152, pp.155-184.
—: "Some Notices on the Magic Agent Wang (Wang *ling-kuan*) at Mt. Ch'i-ch'ü in Tzu-t'ung District, Szechwan Province", in: ZDMG 148, 323-342 (1998)
—: trl., ed.: Leben und Wirken Lao-Tzu's in Schrift und Bild *Lao-chün pa-shih-i hua t'u-shuo*. Würzburg 1990.
—: Der Perlenbeutels aus den Drei Höhlen (*San-tung chu-nang*), Arbeitsmaterialien zum Taoismus der frühen T'ang Zeit, in: AF 112 (1990).
—: Kategorien und Realien im Shang-ch'ing Taoismus, (*Shang-ch'ing tao lei-shih hsiang*), Arbeitsmaterialien zum Taoismus der frühen T'ang Zeit, in: AF 119 (1992).
—: Grundelemente und Tendenzen des religiösen Taoismus, das Spannungsverhältnis von Integration und Individualität in seiner Geschichte zur Chin-,Yüan- und frühen Ming-Zeit, in: MOS 48 (1988).

–: "The Scripture of the Hidden Contracts (*Yin-fu ching*): a short survey on facts and findings", in: Nachrichten der Gesellschaft für Natur- und Völkerkunde Ostasiens 136, pp.75-83 (1984).

–: „Die Ausführungen Li Tao-yüans zur Geschichte und Geographie des Berges Lu (Chiang-hsi) im <Kommentar zum Wasserklassiker>, und ihre Bedeutung für die regionale Geschichtsschreibung, in: *Oriens Extremus* 28 (1981).

–: Der Bericht über den Berg Lu (*Lu-shan chi*) von Ch´en Shun-yü, ein historiographischer Beitrag aus der Sung Zeit zum Kulturraum des Lu Shan. München 1978.

I. Robinet: Méditation taoiste. Paris 1979.

M.Saso: Taoism and the Rite of Cosmic Renewal. Washington 1972.

K.Schipper and F.Verellen eds.: The Taoist Canon, A Historical Companion to the Daozang. 3, vols. Chicago 2004.

K.Schipper: "Le Calendrier de Jade, Note sur le Laozi zhongjing", in: *Nachrichten der Gesellschaft für Natur- und Völkerkunde Ostasiens* 125, pp.75-80.

G. Schlegel: Uranographie Chinoise. Leiden 1875.

A.K.Seidel: La divinisation de Lao Tseu dans le taoisme des Han. Paris 1969.

Shih-chi. Ed.: Tung-hua Comp. Taipei 1970.

L.Skar: "Administering Thunder: A thirteenth Century Memorial Deliberating the Thunder Rites", in: *Cahiers d´Extrême-Asie* 9, pp.159-202. (1996-1997).

– : "Ritual Movements, Deity Cults, and the Transformation of Daoism in Song and Yuan Times", in: L.Kohn ed.: Daoism Handbook, pp.413-463. Leiden 2000.

W.E.Soothill and L. Hodous: A Dictionary of Chinese Buddhist Terms. Rpr. Taipei1972.

Ting Fu-pao: *Fo-hsüeh ta-tz´u-tien*. Taipei 1974

B.Watson trl.: The Complete Works of Chuang Tzu. New York 1968.
Wang Ch´ung: *Lun-heng*. Ed. Shanghai 1974.
Wang Tsung-yü: *Dao-chiao i-shu yen-chiu*. Shanghai 2001.

The Sources in the Taoist Canon

TT 15 *Wu-shang chiu-hsiao yü-ch'ing ta-fan tzu-wei hsüan-tu lei-t'ing yü-ching*

TT 31 *Huang-ti yin-fu ching*

TT 110 *Huang-ti yin-fu ching shu*

TT 147 *Ling-pao wu-liang tu-jen shang-p'in miao-ching fu-t'u*

TT 148 *Wu-liang tu-jen shang-p'in miao-ching p'ang t'ung-t'u*

TT 171 *Ch'ing-wei hsien-p'u*

TT 173 *Chin-lien cheng-tsung chi*

TT 174 *Chin-lien cheng-tsung hsien-yüan hsiang-chuan*

TT 263 *Hsiu-chen shih-shu tsa-chu chih-hsüan p'ien*

TT 295 *Hsü-hsien chuan*

TT 296 *Li-shih chen-hsien t'i-tao t'ung-chien*

TT 599 *Tung-t'ien fu-ti yueh-tu ming-shan chi*

TT 622 *T'ai-shang hsüan-ling pei-tou pen-ming yen-sheng chen-ching*

TT 629 *T'ai-shang pei-tou erh-shih-pa chang ching*

TT 670 *Nan-hua chen-ching*

TT 682 *Tao-te chen-ching chu*

TT 1015 *Chin-so liu-chu yin*

TT 1016 *Chen-kao*

TT 1101 *T'ai-p'ing ching*

TT 1129 *Tao-chiao i-shu*

TT 1130 *Tao-tien lun*

Bibliography

TT 1032 *Yün-chi ch'i-ch'ien*

TT 1139 *San-tung chu-nang*

TT 1166 *Fa-hai i-chu*

TT 1168 *T'ai-shang Lao-chün chung-ching*

TT 1220 *Tao-fa hui-yüan*

TT 1221 *Shang-ch'ing ling-pao ta-fa*

TT 1227 *T'ai-shang chu-kuo chiu-min tsung-chen pi-yao*

TT 1232 *Tao-men shih-kuei*

TT 1241 *Ch'uan-shou san-tung ching-chieh fa-lu lüeh-shuo*

TT 1248 *San-tung ch'ün-hsien-lu*

TT 1250 *Ch'ung-hsü t'ung-miao Shih-ch'en Wang hsien-sheng chia-hua*

TT 1402 *Shang-ch'ing huang-t'ing wu-tsang liu-fu chen-jen yü-chou ching*

TT 1476 *Sou-shen chi*

TT 1485 *Tzu-wei tou-shu*

Glossary

A Corpus of Taoist Ritual *Tao-fa hui-yüan* 道法會元　14, 38, 40, 41, 59, 60, 64-66, 69, 111
A Dictionary of Official Titles in Imperial China　4
Agrarian communities　4
Altar　7, 10, 21, 36, 38, 44, 48, 49, 51, 56, 59, 65, 78, 79, 82, 84, 87, 88, 93, 100-102, 108, 115, 116
Amulet *fu* 符　7, 8, 10, 29, 30, 36, 40-44, 46-48, 55, 56, 59, 64, 65, 71, 73, 78, 79, 86, 88-92, 95, 96, 103, 106, 107, 111, 112, 114, 116, 118, 119
Amulet water *fu-shui* 符水　72, 96
Ancestral breath *tsu-ch'i* 祖氣　22, 24, 42, 44, 61, 63
Ancestral palace *tsu-kung* 祖宮　20
Ancestors　73
Anterior Heaven *hsien-t'ien* 先天　23, 25, 28, 29, 35, 56, 59, 60, 64

Barefooted　65, 115
Black crow　22
Body of a snake　45
Blood　45, 116
Blood sacrifices　73, 83, 115
Breathing techniques *fu-ch'i* 服氣　5, 9, 41, 42, 48, 49, 61, 106
Bright Hall *ming-t'ang* 明堂　48, 51, 52
Buddha　63

Central palace *chung-kung* 中宮　19, 21, 26, 28, 43, 61
Cha-tzu 劄子　95
Chai 齋　9, 10, 13
Chang-shui shih-che 掌水使者　117
Chang Wan-fu 張萬福　3
Chang Yü-ch'u 張宇初　15, 39
Ch'ang-chiang 長江　101

Chao Hsüan-t'an/Kung-ming 趙玄壇公明　20, 84
Chao-lung chih-yü chou 召龍致雨咒　101
Cheng-fa chu-ko jih-chih yüan-chün 正法主格日直元君　117
Cheng-i /Heavenly Master Taoism 正一　3, 13
Cheng/chen-hsing 正/真形　42, 95
Ch'eng-huang 城隍　79, 84, 93, 110
Chi-kao 祭告　83
Ch'i-lung chih-yü fu 起龍致雨符　92, 93
Chi lü-ling ta-shen 祭律令大神　85
Chi-shen chou 祭神咒　86
Ch'i-ch'ing chou 祈晴咒　105
Ch'i-ch'ing t'an-shih 祈晴壇式　102
Ch'i-chuan chou 七轉咒　91
Ch'i-kang chiang-chün 起罡將軍　112
Ch'i-shen 氣神　26
Ch'i-yü chou 祈雨咒　103
Ch'i-yü t'an shih 祈雨壇式　100
Chiao 醮　4, 13, 101, 103
Chien-chüeh 劍訣　21, 30
Chien-kang chiang-chün 建罡將軍　112
Chien-yü fa 建獄法　96
Ch'ih Yu 蚩尤　85
Chin-ching tu-hai chih yü 金精毒害之獄　97
Chin-lien cheng-tsung chi 金蓮正宗記　17
Chin-lien cheng-tsung hsien-yüan hsiang-chuan 金蓮正宗仙源像傳　17
Ching 井　90
Ch'ing-t'ang 清堂　96
Ch'ing-wei 清微　2, 14, 17
Ch'ing-wei hsien-p'u 清微仙譜　17
Chiu-ch'üan k'u-nao chih yu 九泉苦腦之獄　97
Chiu-t'ien chin-ch'üeh yu-shih 九天金闕御史　118
Chou-hsiang t'u-fu fa 咒香吞服法　107

Ch'u-ku chou 出穀咒　36
Ch'u-yin chou 出印咒　80
Chuan-hun chien-sung shih-che 縛魂監送使者　99
Ch'uan-k'o 傳科　76
Ch'uan-ling shih-che 傳令使者　99
Ch'uan-tu 傳度　79, 117
Chung hsiao jen i 忠孝仁義　82
Chü-hsing 聚形　92
Ch'ü-lung hsi-shui kang 驅龍吸水罡　103, 113
Ch'üan-chen 全真　17
Chüan-shui ch'in-lung 捲水擒龍　61
Cinnabar field *tan-t'ien* 丹天　34, 42, 47
Clear sky　7, 48, 51, 78, 86, 88, 95, 100, 102, 105
Colon *ku-tao* 谷/穀道　30, 57, 58, 61
Communal rituals　4
Confucianism　1
Consecration　73, 79, 107
Cosmos　27, 48, 49, 55, 57, 58, 61, 63, 94
Creation　17, 61
Creative impetus　57
Crusade *fa* 伐　82, 109, 110, 115

Delivery (child)　45
Dispersed form, see *san-hsing*
Divination　1
Divine nature　16
Divine patrons *chu-fa* 主法　59
Divine texts *ling-wen* 靈文　73
Divine prefectures *shen-fu* 神府　75
Divine recipies *shen-fang* 神方　73
Door of life *sheng-men* 生門　23
Dragons (and snakes)　73, 77, 78, 85, 94, 102, 103, 110, 111
Dragon-and-tiger amulet *lung-hu fu* 龍虎符　6
Dragon-pools　93, 100

Eight trigrams see *pa-kua* 八卦
Elixirs　78
Elixir classics *tan-ching* 丹經　22
Encyclopaedias　1, 39
Exorcist rituals/exorcism/exorcist　2, 4, 7, 10, 13, 15, 31, 65

Fa-ch'ien wen-tzu chou 發遣文字咒　114
Fang Chung-kao 方仲高　40
Faith healer　13
Fei-chien chan t'ien-huang 飛劍斬天皇　44
Fei-chien chuo-lung fu 飛劍捉龍符　92
Fei-lung chan-kuai kang 飛龍斬怪罡　110, 112
Fei-lung chih-yü kang 飛龍致雨罡　101, 112
Feng-fa 奉法　76
Feng-hou 風后　77
Feng-po ta-shen 風伯大神　102
Fire　46, 52, 55-57, 80, 89, 90, 94, 104, 109, 116
Fire Master　17, 65, 69, 73, 81, 87, 96, 100, 102, 107, 109, 112, 114, 115
Fire of the stars *hsing-huo* 星火　114
Fire official *huo-kuan* 火官　26
Fire-thunder *huo-lei* 火雷　53
Five colours　47
Five elements　19, 21
Five intestines　19, 20, 21, 34
Five Thunders *wu-lei* 五雷　34, 37, 65, 70-74, 77-81, 87, 89, 90, 91, 95, 96, 98, 99, 100, 101, 108, 111, 119
Fox spirits 14
Fu-chüeh 符訣　73
Fu-feng meng-li Hsin t'ien-chün ta-fa 負風猛吏辛天君大法　36
Fu-feng meng-shen 負風猛神　102

Gall *tan* 膽　20, 24, 26, 34, 36
Gate of heaven *t'ien-men* 天門　47, 101
God-emperors　37, 71, 72, 80-82, 85, 88, 89, 94, 95, 100-103, 105, 108, 109, 110, 114, 116, 120
Green face　45, 46
Grind the teeth *k'ou-ch'ih* 叩齒　35, 47, 89, 106, 114

Hades　73, 106, 120
Handle of the dipper, see *T'a-fan tou-ping* 28-30, 61
Han Po-wen 韓伯文　104
Heart/*hsin* 心　19, 20, 24, 26, 29, 34, 45, 46, 48-50, 52, 53, 55-57, 59, 60, 63

Glossary

Heart of heaven *t'ien-hsin* 天心 16, 75
Heavenly documents *t'ien-shu* 天書 6, 10
Heavenly eye *t'ien-mu* 天目 41, 106
Heavenly female *hsüan-p'in* 玄牝 22, 57
Heavenly laws *t'ien-lü* 天律 37
Hells, see spirit hells
Ho-t'u 河圖 50
Hsiang-hao p'in 相好品 39
Hsien-t'ien i-ch'i huo-lei Chang shih-che ch'i-tao ta-fa 先天一氣火雷張使者祈禱大法 60, 64
Hsin-huo 心火 26
Hsin shuai 辛帥 20, 33, 34, 36, 38, 40
Hsing-ming 性命 83
Hsü-hsien chuan 續仙傳 5
Hsü Sun 許遜 5, 9, 77, 95
Hsüan-huang 玄黃 25
Hsüan-kuan 玄關 43
Hua-shen 化身 88, 89
Hua-shen wen 化神文 114
Huang-ti 黃帝 77, 85
Human body 16, 18, 21, 22, 31, 33, 36, 45, 48, 54, 55, 57, 61, 62, 73
Hun- and/or p'o souls *hun-p'o* 魂魄 42, 51, 73
Hun-ming 混明 60
Hung-fan 洪範 27
Huo-chü yüan-shuai 火車元帥 114
Huo-kuang liu-hsing yu 火光流星獄 97
Huo-ling chih chai 火鈴之宅 86
Huo-lun lü-ling 火輪律令 106
Huo-pu shang-shu 火部上書 118

Icons 108
Immortality/immortals 2, 6
Incense burner/table 98, 102, 103, 108, 115
Infant *ying-er*h 嬰兒 32
Internal practice *hsing-ch'ih* 行持 24, 27, 49, 54, 73, 114
Iron 93, 111
I-shih chüeh 役使訣 116
I-shih yin 役使印 113

Jade Emperor *yü-ti* 玉帝 37, 81, 104
Jade Pivot, see *Yü-shu*
Ju-yin chou 入印咒 81

Jui-kuang hsien-shih 瑞光仙師 108

Kalpa 82
Kang-mu 罡目 113
Kang-pu 罡步 88
Kidneys/*shen* 腎 19, 24, 30, 32, 34, 35, 48-50, 52, 54-56, 58, 61
Killing breaths *sha-ch'i* 煞氣 43, 57, 58
K'ou-ch'ih 叩齒, see grind the teeth
Kuei-mu 鬼目 113
K'uei-hsing/kang 魁星/罡 60, 90
K'uei-mu 魁目 113
K'un-lun 崑崙 51, 94
Kung-ts'ao 功曹 79
Kuo Yüan-ching 郭元京 40

Lamp-wick *teng-hsin* [*ts'ao*] 燈心[草] 56, 57
Lei-ch'eng t'ieh-ch'üan 雷城鐵券 119
Lei-ch'i 雷氣 80, 90
Lei-chü 雷局 28, 29, 30, 33-35, 38, 45, 56, 87, 88, 91, 95, 111, 113
Lei-fa pi-chih 雷法秘旨 31
Lei-fa ti-tzu 雷法弟子 117
Lei-ku 雷鼓 95, 99
Lei-kung, see Thunder Lord
Lei-tien k'ao-chih yü 雷電考治獄 97
Lei-t'ing ti-chün 雷霆帝君 89
Lei-t'ing pan-kao 雷霆版誥 118
Lei-t'ing shen-wei 雷霆神位 85
Lei-t'ing tu-ssu kuei-shen kung-shih 雷霆都司鬼神公事 118
Lien-chen 廉貞 26, 60
Lien-t'ien t'ieh-chang kang 連天鐵障罡 111, 113
Liu Fang 劉昉 8, 9
Liu-tan fu 流丹符 93
Liver/*kan* 肝 19, 20, 26, 49, 52, 91
Locusts 72
Lotus flower 48, 50, 52
Lung-chün 龍君 83
Lun-heng 論衡 3
Lung-kung 龍宮 82, 83
Lung-lei 龍雷 71, 81, 82
Lü-ling ta-shen 律令大神 85, 86
Ma Ling-kuan 馬靈官 20
Ma Yü-lin 馬鬱林 40
Mao-chün 茅君 78

Martial spirit forces 2
Memoranda/memorialise 76, 78, 79, 81-83, 88, 98, 100-102, 110
Meng-tzu 孟子 42
Ming-men 命門 42
Ming-mu kuang-tzu fu 明目光字符 106
Mouth 49, 53
Mudrâs 64, 65, 70
Mu-lang chou 木郎咒 104
Mu-lao 木老 26
Mysterious female, see heavenly female

Nan-tou/Dipper of the South 南斗 51
Naval 56
Ni-wan/Mud Pill 泥丸 35, 47, 58
Nei-tan/internal alchemy 內丹 30, 31, 63, 64
Northern Dipper/Big Dipper pei-tou 北斗 29, 47, 52, 59-61, 71, 75, 79, 80, 113
Nostrils 21, 28-30, 41, 42, 43, 44, 56-58, 61, 62

Open air 115
O-po 闕伯 103
Oral transmission k'ou-ch'uan/chüeh 口傳/訣 25, 27, 58, 65, 86, 103
Original breath yüan-ch'i 元氣 22, 42-44

Pa-kua/eight trigrams 八卦 48, 54, 80, 94
Pa-kua lei-shen 八卦雷神 35
Pai Yü-ch'an 白玉蟾 31, 64
Pantheon 65
Pao ming-yüeh shang-ch'ing lu 保明兌上清錄 117
Pei-chi 北極 72
Pei-ti 北帝 75, 89, 99
Perfect King, see Yü-ch'ing chen-wang
P'eng-lai (Office) 蓬萊 72, 75, 89, 99
P'eng-shan chu-hai 烹山煮海 51
Petitons 3, 37, 79, 93, 98
P'iao-hsing 飄星 60
Pien-shen 變神 36
Pien-shen chüeh 變神訣 88, 114
P'o-chün 破軍 59, 60
Possession 36
Posterior Heaven hou-t'ien 後天 35, 36

P'o-ti chao-lei kang 破地召雷罡 88, 95, 115
Promotion 117
Pu-kang pien-shen chou 步罡變神咒 33

Quail 32

Red hair 32, 39, 85
Register lu 錄 9, 10, 36, 38-40, 65, 114
Ritual steps 64, 65, 88, 110-112

San-chieh/three realms 三界 82, 92, 97, 115
San-ching chou 三淨咒 28
San-hsing 散形 92, 111
San-kuan 三官 82, 119
San-wu 三五 105
Sanskrit 36
San-tung chu-nang 三洞珠囊 13, 39
School Talks 45, 64
Seals 71, 76-81, 110, 112, 114, 120
Self tzu-chi 自己 16
Self-cultivation 2, 24, 27, 31, 49, 73, 85, 114
Seven apertures 29
Shaman/mediumistic culture 2, 36
Shang-ch'ing 上清 38, 117
Shang-ch'ing ling-pao ta-fa 上清靈寶大法 46
Shang-ch'ing ssu-ming yü-fu yu-ch'ing 上清司命玉府右卿 118
Shang-ch'ing yü-fu wu-lei shih-yüan 上清玉府五雷使院 119, 120
Shang-ch'ing yü-fu wu-lei ta-fa yü-shu ling-wen 上清玉府五雷大法玉樞靈文 64
Shang-yüan i-p'in t'ien-kuan 上元一品天官 77
Shao-yang lei-kung 劭陽雷公 114
She-ling lei 社令雷 71, 81, 83, 84
Shen Fen 沈汾 5
Shen-hsiao hu-chao chüeh 神霄呼召訣 88
Shen-hsiao tao/t'ien 神霄道/天 2, 69, 81
Shen-lei 神雷 71, 81, 82, 86, 109
Shen-mu 神目 113
Sheng-ch'i ch'ing-lung fu 生氣青龍符 91
Shou kang-tou chou 收罡斗咒 96

Glossary

Shou-shui tso-ch'ing chou 收水作晴咒 103
Shrine/temple 54, 65, 73, 84, 96, 98, 109, 110, 115
Shu-chi 樞機 71
Shu-fu shih 書符式 46
Shu-hsing 樞星 60
Shui-lei 水雷 71, 81, 83, 109
Silver teeth 32, 39, 85, 87, 105
Six-and-One *liu-i* 六一 53
Snow 107, 108, 109
Spell 7, 19, 21, 28, 29, 30, 33, 34, 47, 65, 79, 81, 86-88, 90-94, 101, 104, 105, 107, 109, 114-116
Spine *chia-chi* 夾脊 30, 51, 52, 56
Spirit hells 64, 65, 96, 97
Spiritual/divine forces 18, 22, 24, 27, 28, 35, 36, 42, 44, 55, 56, 61, 63, 78, 106
Spleen/*p'i* 脾 21, 24, 34, 55, 56, 57, 58, 63
Ssu-yüan 四院 76
Ssu-sha 四煞 90
State rituals 14
Stomach 21
Stones 50, 52
Sung Hui-tsung 宋徽宗 4, 14, 48
Sung-period 2, 3, 4, 14, 40, 54
Sweat 55, 57
Sword 45, 101, 103, 108, 111, 114-116

T'a-fan tou-ping 踏翻斗柄 61
Ta-tung ch'ing-yü chou 大洞請雨咒 104
T'ai-i 太乙 60, 61, 93, 104
T'ai-i Cave *t'ai-i hsüeh* 太一 58, 60
T'ai-p'ing ching 太平經 13
T'ai-po feng-ling 太伯風令 104
T'ai-shang Lao-chün 太上老君 17, 107
T'ai-shang Lao-chün chung-ching 太上老君中經 39
T'ai-sui 太歲 100, 108
Taiwan 2, 65
Tan-fu 丹符 100
Taoist Canon 10, 15
Tao-chiao i-shu 道教義樞 13
Tao-miao 道妙 17
Tao-shu 道術 5, 8
Tao-tian lun 道典論 13

Teng Kung-ch'en 鄧拱辰 40
Teng Shuai Po-wen 鄧帥伯溫 19, 26, 32-34, 39, 85, 105
Three eyes 45, 46
Three Heavens *san-t'ien* 三天 72
Three Marshals *san-shuai* 三帥 35
Three Monitoring Offices *san-ssu* 三司 75, 82, 88, 94, 96, 98, 100-102, 108
Three potentials *san-ts'ai* 三才 62
Three Terraces *san-t'ai* 三台 29
Thunder altar *lei-t'an* 雷壇 51, 56
Thunder breaths see *lei-ch'i*
Thunder deities/divinities *lei-shen* 雷神 9, 16, 17, 24, 30, 33, 35, 40, 59, 64, 71, 89, 110
Thunder department *lei-pu* 雷部 50, 85, 95, 100, 103, 115
Thunder drums. See *lei-ku*
Thunder files *lei-pu* 雷簿 39
Thunder Lord/*Lei-kung* 雷公 3, 33, 77, 87, 92, 114
Thunder Magic/Rituals *wu-lei fa* 五雷法 2, 3, 4, 10, 13-18, 31, 33-35, 38, 40, 41, 47, 48, 54, 58, 59, 61, 62, 64-66, 114, 117
Thunder Office, see *lei-chü* 雷局
Thunder prefecture *lei-fu* 雷府 74
Thunder ranks 35
Thunder Wall *lei-ch'eng* 雷城 72, 74, 103, 119
Three worlds *san-chieh* 三界 51
Ti-huo lei-shen 地火雷神 91
Tien-lu k'ao-chin shih che 典錄拷禁使者 99
T'ien-chang 天章 89
T'ien-fu 天符 94, 97
T'ien-hsi 天熹 53
T'ien-hsin cheng-fa 天心正法 2
T'ien-huo lei-shen 天火雷神 90
T'ien-huang fu 天皇符 44, 45
T'ien-i 天一 50
T'ien-kang 天罡 26, 59, 61, 79
T'ien-kuan ch'i-sha fu 天關七煞符 111
T'ien-kuan fu 天關符 110
T'ien-lei 天雷 71, 81, 82
T'ien-ming 天命 102
Tien-mu/mother of lightning 電母 33, 92

T'ien-shu yüan 天樞院 120
T'ien Yüan-tsung 田元宗 40
Tongue 34, 36, 49, 53, 55, 57
Tou-chung tu-shui shih-che 斗中都水使者 117
Tou-chung/hsia t'ung-shih she-jen 斗中/下通事舍人 77, 79
Tou-k'uei yin 斗魁印 78
Tou-shu yüan 斗樞院 75
Tripod vessel 22
Tso-shui chao-lei kang 作水召雷罡 101, 108, 112
Tsung-she yin 總攝印 113
Tu-jen ching 度人經 25
Tu-shui yu/tso p'an-kuan 都水右左判官 117
T'u-ti 土地 110
Tung Feng 董奉 8
Tung-chi pa-ling chou 東極八靈咒 109
Tung-ling chou 東靈咒 108
*Tung-ling shang-hsia*ng 東靈上相 108
Tzu-hu 紫戶 42
Tzu-kuang t'ung-tzu 紫光童子 77
Tzu-wei 紫微 105, 106
Tz'u 詞 82, 100, 102, 108, 109

Urinary organs 61
Urinate/urine *hsiao-i* 小遺 49, 54, 61

Wang Ch'ung 王充 3
Wang Shih-ch'en ch'i-tao pa-tuan chin 王侍宸祈禱八段錦 18
Wang Tzu-hua 汪子華 17
Wang Shan Ling-kuan 王善靈官 20
Wang Wen-ch'ing/Shih-ch'en 王文卿侍宸 14, 15, 17-19, 22, 24, 28, 30, 31, 33, 39-41, 43, 46, 48, 54, 55, 60, 62-64, 66, 69
Water Department *shui-fu* 水府 97, 110
Water monsters *shui-yao* 水妖 78, 83, 89, 91, 96, 97, 102, 110, 111
Wen-shuai 溫帥 20, 26
Windlass pass *lu-lu kuan* 轆轤關 51
Window of Death *ssu-hu* 死戶 23
Window of Earth *ti-hu* 地戶 29, 30, 52, 101, 103
Window of the brain *nao-hu* 腦戶 51, 52, 58
Womb 42, 46, 51
Wu-fang man-lei [*shih che*] 五方蠻雷[使者] 33, 34, 116
Wu-lei chu-tsai 五雷主宰 88
Wu-lei chüeh 五雷訣 95
Wu-lei chou 五雷咒 95
Wu-lei tsung-she fu 五雷總攝符 89
Wu-lei yüan 五雷院 76
Wu-lei yüan yu p'an-kuan 五雷院右判官 117
Wu Meng 吳猛 5, 9, 41, 77, 95
Wu-so fu(-chüeh) 五鎖符(訣) 90
Wu-tang shan 武當山 85
Wu-ti 五帝 114
Wu-yüeh chang-jen 五嶽丈人 71
Wu-yüeh wu-chün chang-jen 五岳五君丈人 77
Wu-shang pi-yao 無上祕要 1
Wu-tao 五道 98

Yeh Ch'ien-shao 葉千劭 4-10, 41, 65
Yellow court *huang-t'ing* 黃庭 25, 28, 43, 56
Yen-huo chen-hsing fu 焱火真形符 103
Yen-huo fu 焱火符 86
Yen-huo ta-shen 焱火大神 85, 102
Yen-huo ta-shen fu 焱火大神符 94, 110
Yu-t'ai ch'ang-yeh chih yü 幽臺長夜之獄 98
Yung-yin chou 用印咒 80
Yü-ch'en tung-ling chih yin 玉晨洞靈之印 78, 110, 111
Yü-ch'ing chen-wang 玉清真王 69, 70, 71, 72, 74, 88
Yü-ch'ing chüeh 玉清訣 47
Yü-feng 御風 18, 19, 22, 24, 27, 29-31, 40, 42, 43, 49, 53, 56, 58, 60-62
Yü-fu shang-ch'ing wu-lei fu shih-p'an 玉府上卿五雷副使判 118
Yü-shu 玉樞 72, 73, 75-77, 88, 89, 99, 119
Yü-shu yüan shih 玉樞院使 119
Yü-wan t'ung-tzu 玉完童子 79
Yü-wei fan-p'o 魚尾翻波 52
Yüan-chün 元君 93

Yüan-ming chen-chün 元命真君 105
Yüan-shih t'ien-tsun/Heavenly Worthy of Prime Origin 元始天尊 25, 46, 47
Yüan Wu-chieh 袁無介 18, 19, 40, 41, 48, 55, 62
Yün-lei ta-sui fu 運雷打祟符 95